Annals of Mathematics Studies

Number 93

SEMINAR ON MICRO-LOCAL ANALYSIS

BY

VICTOR W. GUILLEMIN,
MASAKI KASHIWARA,
AND TAKAHIRO KAWAI

PRINCETON UNIVERSITY PRESS
AND
UNIVERSITY OF TOKYO PRESS

PRINCETON, NEW JERSEY
1979

Published in Japan exclusively by
University of Tokyo Press
In other parts of the world by
Princeton University Press

Printed in the United States of America
by Princeton University Press, Princeton, New Jersey

Library of Congress Cataloging in Publication data will
be found on the last printed page of this book

Math Sci (Sup)

TABLE OF CONTENTS

PREFACE

This book is the outgrowth of a seminar on micro-local analysis sponsored by the Institute for Advanced Study during the academic year 1977-1978. For the benefit of the general reader we will attempt, in a few words, to put the subject matter of this volume into a historical perspective. By micro-local analysis we mean the study of generalized functions as local objects on the cotangent bundle. In a sense micro-local analysis has its roots in the work of Cauchy, Riemann and Hadamard on the relationship between singularities of solutions of partial differential equations and the geometry of their characteristics. However the theory we will be concerned with here really starts about 1970 with Sato's definition of microfunctions as localizations of hyperfunctions and with the work of Maslov, Egorov and Hörmander on quantized contact transformations (or Fourier integral operators). These two closely related developments enabled one to study in much more meticulous detail than was ever before possible the singularities of solutions of partial differential equations and of generalized functions arising naturally in geometric and group-theoretic contexts.

The first series of lectures in this volume are an introductory account of the theory of microfunctions. This parallels somewhat the account in [SKK]; however, here the cohomological aspects of the subject are somewhat suppressed in order to make these lectures more accessible to an audience of analysts. The subsequent lectures in this volume are devoted to special aspects of the theory of microfunctions and to applications such as boundary values of elliptic partial differential equations, propagation of singularities in the vicinity of degenerate characteristics, holonomic systems, Feynman integrals from the hyperfunction point of view and harmonic analysis on Lie groups.

Seminar on Micro-Local Analysis

INTRODUCTION TO THE THEORY OF HYPERFUNCTIONS

Masaki Kashiwara

§0. *Introduction*

The purpose of this note is to give an introduction to the theory of hyperfunctions, microfunctions and micro-differential operators.

Hyperfunctions were introduced by M. Sato (J. Fac. Sci. Univ., Tokyo, Sect. I, 8 (1959), 139-193; 8 (1960), 387-437). For Sato a hyperfunction is a sum of boundary values of holomorphic functions. In order to formulate the theory in a rigorous way, he introduced local cohomology groups and expressed hyperfunctions as cohomology classes. Here, we employ a more intuitive way of defining hyperfunctions. For a rigorous justification of our approach we refer to the article: Sato-Kawai-Kashiwara, Microfunctions and pseudo-differential equations, Lecture Notes in Math. No. 287, Springer, 1973 pp. 265-529 (abbreviated by S. K. K.).

§1. *Hyperfunctions*

1.1. *Tangent Cones.* We will need some geometric preliminaries. We will begin with the definition of *tangent cones.*

Let M be a C^1-manifold. We shall denote by TM the tangent vector bundle, T^*M the cotangent bundle, $\tau : TM \to M$, $\pi : T^*M \to M$ the canonical projections. T_xM is the tangent vector space at $x \in M$ and T_x^*M is the cotangent vector space at $x \in M$.

Seminar on Micro-Local Analysis
0-691-08228-6/79/00 0003-36 $01.80/1 (cloth)
0-691-08232-4/79/00 0003-36 $01.80/1 (paperback)
For copying information, see copyright page

Take a point x in M and a local coordinate system $(x_1, \cdots, x_\ell)$ in a neighborhood of x.

DEFINITION 1.1.1. For two subsets A and B of M, the tangent cone $C_x(A; B)$ is the set of limits of sequences $a_n(x_n - y_n)$ where $a_n > 0$, $x_n \in A$, $y_n \in B$, such that x_n, y_n converge to x. We regard $C_x(A; B)$ as a subset of T_xM. Set $C(A; B) = \bigcup_{x \in M} C_x(A; B)$.

REMARK 1.1.2. This definition does not depend on the choice of coordinate systems.

Tangent cones enjoy the following properties:

a) $C(A; B)$ is a closed cone in TM.

b) $C(A; B) = -C(B; A)$.

c) $C(A; B) = C(\bar{A}; \bar{B})$.

d) $C_x(A; B) = \emptyset \Longleftrightarrow x \notin \bar{A} \cap \bar{B}$.

e) $C_x(A; B) = \{0\} \Longleftrightarrow x$ is an isolated point of $\bar{A}$ and $\bar{B}$.

f) Let $f: M \to N$ be a C^1-map, $y = f(x)$. Then

$$(df)_x C_x(A; B) \subset C_y(fA; fB).$$

If $C_x(A; B) \cap (df)_x^{-1}(0) \subset \{0\}$, then $(df)_x C_x(A; B) = C_y(f(A \cap U); f(B \cap U))$ for a neighborhood U of x.

g) If N is a submanifold of M, then $C_x(A; N) + T_xN = C_x(A; N)$ for $x \in N$.

Notation: We set $C_N(A)_x = C_x(A; N)/T_xN \subset (T_NM)_x = T_xM/T_xN$ and $C_N(A) = \bigcup_{x \in N} C_N(A)_x \subset T_NM$.

1.2. *Definition of* $\tilde{\mathcal{A}}$. Let M be a real analytic manifold (say an open set of $\mathbf{R}^n$), X its complexification (say an open set of $\mathbf{C}^n$), $\mathcal{O}$ the sheaf of holomorphic functions on X, and $\mathcal{A}$ the sheaf of real analytic functions on M.

For $x \in M$, the complex tangent space to X at x can be decomposed as:

$$T_xX = T_xM \oplus \sqrt{-1}\, T_xM ,$$

so we can identify $(T_MX)_x = T_xX/T_xM$ with $\sqrt{-1}\, T_xM$. We shall denote by τ the projection of $\sqrt{-1}\, TM$ onto M.

DEFINITION 1.2.1. For a point $(x_0, \sqrt{-1}\nu_0) \in \sqrt{-1}\, TM$, we say that an open set U of X is an infinitesimal neighborhood of $(x_0, \sqrt{-1}\nu_0)$ if $C(X-U)$ does not contain $(x_0, \sqrt{-1}\nu_0)$ ($\Longleftrightarrow U \ni x + \sqrt{-1}\, t\nu$ for $|x-x_0| \ll 1$, $0 < t \ll 1$ and $|\nu-\nu_0| \ll 1$). For an open cone Ω of $\sqrt{-1}\, TM$, we say that an open set U of X is an infinitesimal neighborhood of Ω if U is an infinitesimal neighborhood of any point in Ω.

DEFINITION 1.2.2. (S.K.K. Def. 1.3.3, p. 276). For an open cone Ω in $\sqrt{-1}\, TM$, we set

$$\tilde{\mathfrak{A}}(\Omega) = \varinjlim_{U} \mathcal{O}(U)$$

where U runs over a set of infinitesimal neighborhoods of Ω and $\mathcal{O}(U)$ is the set of holomorphic functions defined on U.

REMARK 1.2.3. An infinitesimal neighborhood of $(x; \sqrt{-1}\, 0)$ is nothing but a neighborhood of x. Therefore, $\tilde{\mathfrak{A}}(\sqrt{-1}\, TM) = \mathfrak{A}(M)$.

REMARK 1.2.4. (S.K.K. Prop. 1.5.4, p. 285). Suppose that Ω has connected fibers (i.e., $\Omega \cap \sqrt{-1}\, T_xM$ is connected for any $x \in M$). Then $\tilde{\mathfrak{A}}(\Omega) = \tilde{\mathfrak{A}}$ (the convex hull of Ω). Here the convex hull of Ω is the union of the convex hulls of $\Omega \cap \sqrt{-1}\, T_xM$ $(x \in M)$.

1.3. *Definition of hyperfunctions.* Although hyperfunctions are defined by the use of local cohomology in [S], [S.K.K.], we shall give here a more intuitive definition.

Let V be an open set of M. We denote by $\mathfrak{F}(V)$ the totality of the following data: $\{\Omega_i, u_i\}_{i \in I}$ where I is a finite set, Ω_i an open convex cone in $\sqrt{-1}\, TM$ such that $\tau(\Omega_i) \supset V$, and $u_i \in \tilde{\mathfrak{A}}(\Omega_i)$. Let $\sim$ be the equivalence relation on $\mathfrak{F}(V)$ generated by the relation

$\tilde{R} : \{\Omega_i; u_i\}_{i\in I} \; \tilde{R} \{\Omega'_j; u'_j\}_{j\in I'}$ if there are open convex cones Ω_{ij} $(i \in I,\ j \in I')$ and $w_{ij} \in \tilde{\mathcal{O}}(\Omega_{ij})$ satisfying the properties (i) and (ii):

(i) $\Omega_{ij} \supset \Omega_i \cup \Omega'_j$ for $i \in I$ and $j \in I'$.

(ii) $u_i = \sum_{j\in I'} w_{ij},\ v_j = \sum_{i\in I} w_{ij}$.

DEFINITION 1.3.1. We define

$$\mathcal{B}(V) = \mathcal{F}(V)/\sim$$

and call a member of $\mathcal{B}(V)$ a hyperfunction defined on V.

We have the following fundamental properties of $\mathcal{B}$.

THEOREM 1.3.2. (a) $\mathcal{B} : V \mapsto \mathcal{B}(V)$ *is a sheaf.*

(b) *The sheaf* $\mathcal{B}$ *is a flabby sheaf* (*i.e.,* $\mathcal{B}(M) \to \mathcal{B}(V)$ *is surjective for any open set* V).

Let Ω be an open convex cone of $\sqrt{-1}\,TM$. Then, for any $u \in \tilde{\mathcal{O}}(\Omega)$ $(\Omega; u)$ is a member of $\mathcal{F}(V)$, so that we can define the map

$$b_\Omega : \tilde{\mathcal{O}}(\Omega) \to \mathcal{B}(\tau\Omega).$$

THEOREM 1.3.3. (S.K.K. Th. 1.5.2, second row of (1.5.2), p. 283). $b_\Omega : \mathcal{O}(\Omega) \to \mathcal{B}(\tau\Omega)$ *is injective.*

If we regard $\mathcal{B}(V)$ as a space of generalized functions, then b_Ω is the map assigning to each holomorphic function its boundary values. By this notation, $\{\Omega_i; u_i\}_{i\in I} \in \mathcal{F}(V)$ corresponds to the hyperfunction $\sum_{i\in I} b_{\Omega_i}(u_i)$.

Let P be a differential operator with real analytic coefficient, and $\tilde{P}$ its prolongation to a differential operator with holomorphic coefficients.

Then, we define an operation of P on hyperfunction by

$$P\left(\sum b_{\Omega_i}(u_i)\right) = \sum b_{\Omega_i}(\tilde{P}u_i) .$$

REMARK. If Ω_1 and Ω_2 are two convex cones in $\sqrt{-1}\, TM$ such that $\Omega_1 \supset \Omega_2$, then we have

$$b_{\Omega_1}(u) = b_{\Omega_2}(u)$$

for $u \in \tilde{\mathfrak{A}}(\Omega_1)$.

1.4. EXAMPLES.

(a) The δ-function. In this example $M = \mathbf{R}^n$, $X = \mathbf{C}^n$. Define 2^n open sets

$U_{\varepsilon_1,\cdots,\varepsilon_n}$ by $U_{\varepsilon_1,\cdots,\varepsilon_n} = \{z \in \mathbf{C}^n; \varepsilon_j \operatorname{Im} z_j > 0 \text{ for } j = 1,\cdots,n\}$ where $\varepsilon_j = \pm 1$.

Then $U_{\varepsilon_1,\cdots,\varepsilon_n}$ is an infinitesimal neighborhood of $\Omega_{\varepsilon_1,\cdots,\varepsilon_n} = \{(x, \sqrt{-1}\nu); \varepsilon_j \nu_j > 0\}$. Set $u(z) = 1/z_1 \cdots z_n$. Then $u(z) \in \mathcal{O}(U_{\varepsilon_1,\cdots,\varepsilon_n})$. We define the δ-function supported at the origin by

$$\delta(x) = \frac{1}{(-2\pi\sqrt{-1})^n} \sum^{(2^n)} \varepsilon_1 \cdots \varepsilon_n\, b_{\Omega_{\varepsilon_1,\cdots,\varepsilon_n}}(u) . \tag{1.4.1}$$

We will show that $x_j \delta(x) = 0$. We can assume $j = 1$. Set

$$U_{\varepsilon_2,\cdots,\varepsilon_n} = \{z;\ \varepsilon_j \operatorname{Im} z_j > 0\ (j = 2,\cdots,n)\} .$$

Then

$$z_1 u \in \mathcal{O}(U_{\varepsilon_2,\cdots,\varepsilon_n}) .$$

Thus

$$\begin{aligned} x_1\delta(x) &= (-2\pi\sqrt{-1})^{-n} \sum \varepsilon_1 \cdots \varepsilon_n\, b_{\Omega_{\varepsilon_1,\cdots,\varepsilon_n}}(z_1 u) \\ &= (-2\pi\sqrt{-1})^n \sum \varepsilon_1 \cdots \varepsilon_n\, b_{U_{\varepsilon_2,\cdots,\varepsilon_n}}(z_1 u) = 0 . \end{aligned}$$ Q.E.D.

In particular, we have $\operatorname{supp}\delta(x) \subset \{0\}$ (i.e., $\delta(x) = 0$ on $\mathbf{R}^n - \{0\}$). Also we have

$$\left(\sum x_j \frac{\partial}{\partial x_j}\right)\delta(x) = -n\delta(x)\ ,$$

because

$$\left(\sum x_j \frac{\partial}{\partial x_j}\right)\delta(x) = \left(\sum \frac{\partial}{\partial x_j} x_j - n\right)\delta(x) = -n\delta(x)\ .$$

(b) We give another definition of the δ-function.

Let $\{\xi_1,\cdots,\xi_n\}$ be a set of linearly independent real n-vectors. Set $\xi_0 = -\xi_1 - \cdots - \xi_n$. We set $<\xi_j, z> = \sum_{\nu=1}^{n} (\xi_j)_\nu z_\nu$, where $(\xi_j)_\nu$ is the ν-th component of ξ_j.

Set $\Omega_j = \{z \in \mathbf{C}^n;\ \operatorname{Im} < z, \xi_k > > 0 \text{ for } \forall k \neq j\}$. Then, we have

$$(1.4.2)\quad \delta(x) = \frac{1}{(-2\pi\sqrt{-1})^n} \sum_{j=0}^{n} |\xi_0 \wedge \cdots \wedge \widehat{\xi_j} \wedge \cdots \wedge \xi_n|\, b_{\Omega_j}\left(\frac{1}{\prod_{k\neq j} < z, \xi_k >}\right).$$

Here $|\xi_1 \wedge \cdots \wedge \xi_n|$ signifies the absolute value of the determinant of the $n \times n$ matrix $(\xi_1,\cdots,\xi_n)$.

We shall prove that this definition coincides with the one given in Example (a).

Recall the following formula (Feynman's formula):

$$(1.4.3)\qquad (n-1)! \int_{\eta_1,\cdots,\eta_n \geq 0} \frac{\omega(\eta)}{< x, \eta >^n} = \frac{1}{x_1 \cdots x_n}$$

for $x_1,\cdots,x_n > 0$. Here the integral is over the sphere S^{n-1} defined by $\eta_1,\cdots,\eta_n \geq 0$, and $\omega(\eta)$ is the volume element

$$\eta_1 d\eta_2 \wedge \cdots \wedge d\eta_n - \eta_2 d\eta_1 \wedge d\eta_3 \wedge \cdots \wedge d\eta_n + \cdots + (-1)^{n-1}\eta_n d\eta_1 \wedge \cdots \wedge d\eta_{n-1}\ .$$

By a change of coordinate, we have

$$(n-1)! \int_{G(\xi_1,\cdots,\xi_n)} \frac{\omega(\eta)}{<x,\eta>^n} = \frac{|\xi_1 \wedge \cdots \wedge \xi_n|}{<x,\xi_1>\cdots<x,\xi_n>} .$$

Here $G(\xi_1,\cdots,\xi_n)$ is the closed convex cone $\{\eta; \eta = \sum t_j \xi_j; t_j \geq 0\}$.

Let G be an arbitrary closed convex cone. Then

$$u_G(z) = (n-1)! \int_G \frac{\omega(\eta)}{<z,\eta>^n}$$

is holomorphic on $T(G) = \{z \in \mathbf{C}^n; <\mathrm{Im}\, z,\eta> > 0 \text{ for any } \eta \in G-\{0\}\}$. In fact, $<z,\eta> \neq 0$ for $\eta \in G-\{0\}$ and $z \in T(G)$.

Let us consider the decomposition

$$\mathbf{R}^n = \bigcup_j G_j$$

with $G_j \cap G_k$ of measure 0 for $j \neq k$, and consider the hyperfunction $v = \sum b_{T(G_j)}(u_{G_j})$.

Claim. v does not depend on the choice of the above decomposition.

Proof. Let $\mathbf{R}^n = \bigcup_j G_j = \bigcup_k G'_k$ be two such decompositions. $u_{G_j \cap G'_k}$ is defined on $T(G_j \cap G'_k)$ which contains $T(G_j) \cup T(G'_k)$. Therefore,

$$\sum_j b_{(G_j)}(u_{G_j}) = \sum_j b_{T(G_j)}\left(\sum_k u_{G_j \cap G'_k}\right)$$

$$= \sum_j \sum_k b_{T(G_j \cap G'_k)}(u_{G_j \cap G'_k})$$

$$= \sum_k b_{T(G'_k)}\left(\sum_j u_{G_j \cap G'_k}\right)$$

$$= \sum_k b_{T(G'_k)}(u_{G'_k}) .$$

Q.E.D.

Set $G(\varepsilon_1,\cdots,\varepsilon_n) = \{\eta;\ \varepsilon_j\eta_j \geq 0\ (j=1,\cdots,n)\}$. Then

$$u_{G(\varepsilon_1,\cdots,\varepsilon_n)} = \frac{\varepsilon_1\cdots\varepsilon_n}{z_1\cdots z_n}.$$

Thus

$$\sum b_{T(G(\varepsilon_1,\cdots,\varepsilon_n))}(u_{G_{\varepsilon_1,\cdots,\varepsilon_n}}) = \sum \varepsilon_1\cdots\varepsilon_n\, b_{\Omega_{\varepsilon_1,\cdots,\varepsilon_n}}(1/z_1\cdots z_n).$$

If we take the decomposition

$$\mathbf{R}^n = \bigcup_{j=0}^{n} G_j$$

$$G_j = \left\{\eta;\ \eta = \sum_{k\neq j} t_k\xi_k,\ t_k \geq 0\right\},$$

then

$$\sum b_{T(G_j)}(u_{G_j}) = \sum b_{T(G_j)}\left(\frac{|\xi_0\wedge\cdots\wedge\widehat{\xi_j}\wedge\cdots\wedge\xi_n|}{\prod_{k\neq j}\langle z,\xi_k\rangle}\right).$$

(c) In this example $M = \mathbf{R}$. Let $\Omega_\pm = \{z;\ \operatorname{Im} z \gtrless 0\}$, and define

$$(x \pm i0)^\lambda = b_{\Omega_\pm}(z^\lambda),$$

and

$$(x\pm i0)^\lambda(\log(x\pm i0))^m = b_{\Omega_\pm}(z^\lambda(\log z)^m),$$

where, for z^λ and $z^\lambda(\log z)^m$, we take branches on $\mathbf{C} - \{x;\ x \leq 0\}$ such that $z^\lambda|_{z=1} = 1$, $z^\lambda(\log z)^m = \left(\frac{\partial}{\partial\lambda}\right)^m z^\lambda$. We also define

$$x_+^\lambda = [e^{-\pi i\lambda}(x+i0)^\lambda - e^{\pi i\lambda}(x-i0)^\lambda]/(e^{-\pi i\lambda} - e^{\pi i\lambda})$$

$$x_-^\lambda = [(x+i0)^\lambda - (x-i0)^\lambda]/(e^{\pi i\lambda} - e^{-\pi i\lambda}),$$

and

$$Y(x) = [2\pi i - \log(x+i0) + \log(x-i0)]/2\pi i \;.$$

(d) $M = \mathbf{R}^n$, $X = \mathbf{C}^n$. Let $f(x)$ be a (complex valued) real analytic function defined in a neighborhood of $x_0 \in M$. Suppose that $df(x_0) = \vartheta \neq 0$ is a real covector, and that $\mathrm{Re} f(x) = 0$ $(x \in M)$ implies $\mathrm{Im}\, f(x) \geq 0$. Set $\Omega_\varepsilon = \{z \in X;\ \mathrm{Im}\, f(z) > -\varepsilon|\mathrm{Re}\, f(z)|\}$ for $\varepsilon > 0$. Then, for any $\varepsilon > 0$, Ω_ε is an infinitesimal neighborhood of $(x_0, \sqrt{-1}\,v)$ for any $v \in T_{x_0}M$ such that $\langle v, \vartheta\rangle > 0$ (S.K.K. Lemma 3.1.5, p. 306). Therefore, $b_{\Omega_\varepsilon}(f^\lambda)$ is well defined, (i.e., boundary value from the direction of $\mathrm{Im}\, f > 0$). We shall denote this hyperfunction by $(f+i0)^\lambda$.

(e) Set $\Omega_\pm = \{z \in \mathbf{C}^n;\ \pm\mathrm{Im}\, z_1 > \sqrt{|\mathrm{Im}\, z_2| + \cdots + |\mathrm{Im}\, z_n|}$ and $f(z) = z_1^2 - z_2^2 - \cdots - z_n^2$. Then $f(z) \neq 0$ on $\Omega_\pm$. Therefore, we can define

$$b_{\Omega_\pm}(f^\lambda)$$

which we shall denote by $((x_1 \pm i0)^2 - x_2^2 - \cdots - x_n^2)^\lambda$.

(f) $M = \mathbf{R}^2$, $X = \mathbf{C}^2$, $M \ni (x, y)$, $X \ni (z, w)$, $z|_M = x$, $w|_M = y$.

Set $\Omega = \{(z, w) \in X;\ \mathrm{Im}\, z, \mathrm{Im}\, w > 0\}$, $f(z, w) = z + w^{2/3}$, where we take a branch of $w^{2/3}$ such that $(\sqrt{-1})^{2/3} = e^{\pi\sqrt{-1}/3}$. Then $f(z, w)$ is holomorphic on Ω and never vanishes on Ω. Therefore, f^λ is also holomorphic, and $b_\Omega(f^\lambda)$ is well-defined. We note that

$$\left(\frac{\partial}{\partial x}\right)^m b_\Omega(f^\lambda) = \lambda(\lambda-1)\cdots(\lambda-m+1)\, b_\Omega(f^{\lambda-m}) \;.$$

1.5. *Relations with distributions.* We will show how distributions on a real analytic manifold can be regarded as hyperfunctions.

Let $u(x)$ be a compact supported distribution on $\mathbf{R}^n$. Set

$$\phi(z) = \left\langle u(x), \frac{1}{\prod_{j=1}^{n}(z_j - x_j)} \right\rangle .$$

Then $\phi(z)$ is holomorphic on $\operatorname{Im} z_j \neq 0$.

We associate u with the hyperfunction

$$\frac{1}{(-2\pi\sqrt{-1})^n} \sum^{(2^n)} \varepsilon_1 \cdots \varepsilon_n \, b_{\Omega_{\varepsilon_1,\ldots,\varepsilon_n}}(\phi)$$

where $\Omega_{\varepsilon_1,\ldots,\varepsilon_n} = \{z \in \mathbf{C}^n;\ \varepsilon_j \operatorname{Im} z_j > 0\}$, $\varepsilon_j = \pm 1$. This extends to a homomorphism of sheaves $\mathcal{D}' \to \mathcal{B}$, $\mathcal{D}'$ being the sheaf of distributions. This is an injective homomorphism.

A function $u(z) \in \tilde{\mathcal{A}}(\Omega)$ is called of *polynomial growth* if $u(z)$ satisfies $|\operatorname{Im} z|^N |u(z)| <$ const. Then $u(x+\sqrt{-1}\,ty)$, $t \downarrow 0$, converges in the sense of distribution to $b_\Omega(u)$ (Komatsu: Relative cohomology of sheaves of solutions of differential equations, Lecture Notes in Math., 287, p. 226, 1971; A. Martneau: Distributions et valeurs au bord des fonctions holomorphes, Instituto Gulbenkian de Sciencia, Lisbonne, 1964).

§2. *Microfunctions*

2.1. *Singular Spectrum.* Let $\sqrt{-1}\,T^*M$ be the dual vector bundle of $\sqrt{-1}\,TM$. We will identify $\sqrt{-1}\,T^*M$ with the kernel $T^*X|_M \to T^*M$.

Take a point $(x_0, \sqrt{-1}\,\xi_0) \in \sqrt{-1}\,T^*M$ $(\xi_0 \in T^*_{x_0}M)$.

DEFINITION 2.1.1. A hyperfunction $u(x)$ is called micro-analytic at $(x_0, \sqrt{-1}\,\xi_0)$ if there are open convex cones $\{\Omega_j\}$ and $u_j \in \tilde{\mathcal{A}}(\Omega_j)$ such that $u = \sum_j b_{\Omega_j}(u_j)$ in a neighborhood of x_0 and that $\langle \sqrt{-1}\,v, \sqrt{-1}\,\xi_0 \rangle = -\langle v, \xi_0 \rangle > 0$ for any $v \in \Omega_j \cap \sqrt{-1}\,T_{x_0}M$.

REMARK 2.1.2. u is micro-analytic at $(x_0, \sqrt{-1}\,0)$ if and only if u is zero in a neighborhood of 0.

DEFINITION 2.1.3. Let u be a hyperfunction. Then, we call the singular spectrum of u and denote by $SS(u)$ the set of points of $\sqrt{-1}\,T^*M$ where u is not micro-analytic.

REMARK 2.1.4. In S.K.K., we use slightly different notation from this. There we define the singular spectrum of u to be $(SS(u)-\sqrt{-1}\,T_M^*M)/R^+$, regarded as a subset of the cotangent sphere bundle $\sqrt{-1}\,S^*M$.

Note that SS(u) is a closed cone in $\sqrt{-1}\,T^*M$.

THEOREM 2.1.5. (S.K.K. Prop. 1.5.4, p. 285). *Let Z be a closed cone in $\sqrt{-1}\,T^*M$ containing the zero section $\sqrt{-1}\,T_M^*M$, and $\Omega = \{(x,\sqrt{-1}v) \in \sqrt{-1}\,TM; \langle v,\xi\rangle \geq 0$ for any $(x,\sqrt{-1}\,\xi) \in Z\}$. Suppose that Z is proper convex (i.e., $Z \cap \sqrt{-1}\,T_x^*M$ is a closed proper convex cone for any $x \in M$). Then, for any $u \in \mathcal{B}(M)$, $SS(u) \subset Z$ if and only if u is in $b_\Omega(\tilde{\mathcal{A}}(\Omega))$.*

REMARK 2.1.6. If we take $Z = \sqrt{-1}\,T_M^*M$ in Theorem 2.1.5, we get the following result: *A hyperfunction u is real analytic if and only if* $SS(u) \subset \sqrt{-1}\,T_M^*M$ (S.K.K. Prop. 1.3.5, p. 277). Here $\sqrt{-1}\,T_M^*M$ is by the definition the zero section of $\sqrt{-1}\,T^*M$.

THEOREM 2.1.7. (S.K.K. Cor. 2.1.5, p. 473). (a) *Let u be a hyperfunction and $\{Z_j\}$ a finite set of closed cones in $\sqrt{-1}\,T^*M$ containing $\sqrt{-1}\,T_M^*M$. If $\cup Z_j \supset SS(u)$, then there are hyperfunctions u_j such that $u = \sum u_j$ and $SSu_j \subset Z_j$.*

(b) *Let $\{Z_j\}_{j\in I}$ be a finite set of closed cones in $\sqrt{-1}\,T^*M$ such that Z_j contains $\sqrt{-1}\,T_M^*M$. Let $\{u_j\}_{j\in I}$ be a set of hyperfunctions such that $SS(u_j) \subset Z_j$, and $\sum_{j\in I} u_j = 0$. Then there are hyperfunctions $\omega_{jk}\,(j,k \in I)$ such that*

(i) $\omega_{jk} = -\omega_{kj}$.

(ii) $u_j = \sum_k \omega_{jk}$.

(iii) $SS(\omega_{jk}) \subset Z_j \cap Z_k$.

2.2. *Definition of microfunctions*

Consider the presheaf

$$V \mapsto \mathcal{B}(M)/\{u \in \mathcal{B}(M);\ SS(u) \cap V = \emptyset\}$$

on $\sqrt{-1}\,T^*M$. We denote by $\mathcal{C}$ the sheaf associated with this presheaf.

THEOREM 2.2.1. (S.K.K. Def. 1.3.3, p. 276, and Th. 1.5.3, p. 284). *For any open cone* V *of* $\sqrt{-1}\,T^*M$,

$$\mathcal{C}(V) = \mathcal{B}(M)/\{u \in \mathcal{B}(M);\ SS(u) \cap V = \emptyset\}.$$

We will denote by $sp : \mathcal{B}(W) \to \mathcal{C}(\pi^{-1}W)$ the obvious quotient map.

REMARK 2.2.2. (a) We have

$$\mathcal{C}_{(x,\sqrt{-1}\,\xi)} = \mathcal{C}_{(x,\sqrt{-1}\,t\xi)}, \qquad t > 0,$$

i.e., $\mathcal{C}$ is a constant sheaf along the orbit of $\mathbf{R}^+$ on $\sqrt{-1}\,T^*M - \sqrt{-1}\,T^*_M M$.

(b) $\mathcal{C}_{(x,\sqrt{-1}\,0)} = \mathcal{B}_x$.

N.B. As mentioned before, we will consider $\mathcal{C}$ as living on the vector bundle $\sqrt{-1}\,T^*M$, rather than on the sphere bundle $\sqrt{-1}\,S^*M$ as in S.K.K.

2.3. EXAMPLES. (a) $SS(\delta(x)) = \{(x;\ \sqrt{-1}\xi) \in \sqrt{-1}\,T^*\mathbf{R}^n;\ x = 0\}$. The inclusion $\subset$ is clear. The inclusion the other way will be proved later for $n \geq 2$.

For the moment we will just show that $SS\delta(x) = \{(0;\ \sqrt{-1}\xi)\}$ when $n = 1$. In fact, if $SS\delta(x) \not\ni (0, \sqrt{-1})$, then $sp(x+\sqrt{-1}\,0)^{-1} = (-2\pi\sqrt{-1})\,sp\delta(x)$ at $(0, \sqrt{-1})$, and hence $(x+\sqrt{-1}\,0)^{-1}$ is micro-analytic at $(0, \sqrt{-1})$. Since $SS(x+\sqrt{-1}\,0)^{-1} \subset \{(x, \sqrt{-1}\xi);\ \xi \geq 0\}$, $SS(x+\sqrt{-1}\,0)^{-1} \subset \{(x;\ \sqrt{-1}\xi);\ \xi = 0\}$, and hence $(x+\sqrt{-1}\,0)^{-1}$ is real analytic. This is a contradiction.

(b) $$\begin{aligned} SS(x \pm \sqrt{-1}\,0)^{\lambda} &= \{(x, \sqrt{-1}\xi);\ x\xi = 0, \pm\xi \geq 0\} \quad \text{for } \lambda \neq 0, 1, 2, \cdots \\ &= \{(x, \sqrt{-1}\xi);\ \xi = 0\} \quad \text{for } \lambda = 0, 1, 2, \cdots. \end{aligned}$$

In fact, if $\lambda = 0, 1, 2, \cdots$, $(x \pm \sqrt{-1}\,0)^{\lambda}$ is real analytic and the result is obvious. If $\lambda \neq 0, 1, 2, \cdots$, then $SS(x \pm i0)^{\lambda} \subset \{(x, \sqrt{-1}\xi);\ \pm\xi \geq 0\}$, by

Theorem 2.1.5. Since $(x \pm i0)^{\lambda}$ is real analytic on $x \neq 0$ and not real analytic, $SS(x \pm i0)^{\lambda}$ must be as indicated. (See Remark 2.1.6.)

§3. *Products, pull-back and integration of microfunctions*

In this section we will show that microfunctions have nice analytic properties—e.g., we will show that they can be multiplied, integrated, etc., ... in other words, we will show that analysis on real analytic manifolds can be based on the theory of microfunctions.

3.1. *Proper maps.* A continuous map $f: X \to Y$ from a topological space X to a topological space Y is called *proper* if the pre-image of every point is a compact set and if the image of a closed set of X is a closed set of Y. In particular, when Y is locally compact, f is proper if and only if the inverse image of a compact set of Y is compact. Suppose that X and Y are locally compact. Let Z be a locally closed subset of X, y a point in Y. If $Z \cap f^{-1}(y)$ is a compact set, then there are open neighborhood U of $Z \cap f^{-1}(y)$ and an open neighborhood V of y such that $f(U) \subset V$ and that $U \cap Z \to V$ is a proper map. We note the following lemma:

LEMMA 3.1.1. *Let* X *be a topological space,* $E \to X$, $F \to X$ *two vector bundles and* $f: E \to F$ *a bundle map. Let* Z *be a closed cone of* E. *In order that* $Z \to F$ *be a proper map, it is necessary and sufficient that* Z *does not contain any point* p *in* E *such that* $f(p) = 0$.

3.2. *Products of microfunctions.* Let $u(x)$ and $v(x)$ be two hyperfunctions on M. Suppose that

$$SS(u) \cap SS(v)^a \subset \sqrt{-1}\, T^*_M M \tag{3.2.1}$$

where $SS(v)^a = \{(x, -\sqrt{-1}\xi);\ (x, \sqrt{-1}\xi) \in SS(v)\}$.

THEOREM 3.2.1. (S.K.K. Cor. 2.4.2, p. 297). *Under the condition* (3.2.1),

the product $u(x)v(x)$ *is well defined and* $SS(u(x)v(x)) \subset SS(u) \underset{M}{+} SS(v)$.
Here $A \underset{M}{+} B = \{(x, \sqrt{-1}(\xi_1 + \xi_2)); (x, \sqrt{-1}\xi_1) \in A, (x, \sqrt{-1}\xi_2) \in B\}$.

Proof. If u or v is real analytic, the product is well defined. If not, we define $u(x)v(x)$ in a neighborhood of a point x_0 in M as follows: The condition (3.2.1) implies that there are proper closed convex cones $\{Z_j\}$ and $\{Z'_k\}$ (which contain $\sqrt{-1}\, T_M^*M$) such that

$$\left.\begin{array}{l} SS(u) \subset \cup Z_j \\ SS(v) \subset \cup Z'_k \end{array}\right. \quad \text{in a neighborhood of } x_0 ,$$

and

$$Z_j \cap Z'^a_k \subset \sqrt{-1}\, T_M^*M . \tag{3.2.2}$$

Set $\Omega_j = \{(x, \sqrt{-1}v) : \langle v, \xi \rangle > 0$ for $\forall (x, \sqrt{-1}\xi) \in Z_j\}$, $\Omega'_k = \{(x, \sqrt{-1}v) : \langle v, \xi \rangle > 0$ for $\forall (x, \sqrt{-1}\xi) \in Z'_k\}$. Then, (3.2.2) is equivalent to

$$\Omega_j \cap \Omega'_k \neq \emptyset . \tag{3.2.3}$$

By Theorem 2.1.7, we can write $u = \sum u_j$ with $SS(u_j) \subset Z_j$ and $v = \sum v_k$ with $SS(v_k) \subset Z'_k$. By Theorem 2.1.5, we can represent u_j and v_k as a boundary value of a holomorphic functions; i.e. there are $\phi_j \in \tilde{\mathcal{A}}(\Omega_j)$ and $\psi_k \in \tilde{\mathcal{A}}(\Omega'_k)$ such that $b_{\Omega_j}(\phi_j) = u_j$ and $b_{\Omega'_k}(\psi_k) = v_k$. We shall define uv by

$$\sum b_{\Omega_j \cap \Omega'_k}(\phi_j \psi_k) .$$

The condition (3.2.3) assures us that this is well defined. Using (b) in Theorem 2.1.7, it is easy to check that this definition does not depend on the choice of Ω_j, Ω'_k, u_j, v_k.

Since the singular spectrum of $b_{\Omega_j \cap \Omega'_k}(\tilde{\mathcal{A}}(\Omega_j \cap \Omega'_k))$ is contained in the polar set of $\Omega_j \cap \Omega'_k$, which equals $Z_j \underset{M}{+} Z'_k$, by Theorem 2.1.5,

$$SS(uv) \subset \bigcup_{j,k} (Z_j \underset{M}{+} Z'_k) = \Big(\bigcup_j Z_j\Big) \underset{M}{+} \Big(\bigcup_k Z'_k\Big) .$$

Since this inclusion holds for any choice of $\{Z_j\}$ and $\{Z'_k\}$, we have

$$SS(uv) \subset SS(u) \underset{M}{+} SS(v) . \qquad \text{Q.E.D.}$$

THEOREM 3.2.2. *Let* $u_j(x)$ *be a microfunction defined on an open set* Ω_j *of* $\sqrt{-1}\, T^*M$ $(j=1,2)$. *Let* Ω *be an open set of* $\sqrt{-1}\, T^*M$. *Let us denote by* p *the map from* $\sqrt{-1}\, T^*M \underset{M}{\times} \sqrt{-1}\, T^*M$ *onto* $\sqrt{-1}\, T^*M$ *defined by* $((x,\sqrt{-1}\xi_1), (x,\sqrt{-1}\xi_2)) \mapsto (x, \sqrt{-1}(\xi_1+\xi_2))$. *Suppose that*

$$p^{-1}(\Omega) \cap \operatorname{supp} u_1 \underset{M}{\times} \operatorname{supp} u_2 \to \Omega$$

is a proper map. Then we can define canonically a product $u_1(x)u_2(x)$ *on* Ω *(and* $\operatorname{supp} u_1u_2 \subset p(\operatorname{supp} u_1 \underset{M}{\times} \operatorname{supp} u_2)$*).*

Proof. For the sake of simplicity, suppose that Ω_1 and Ω_2 are cones. Then, $\operatorname{supp} u_1$ and $\operatorname{supp} u_2$ are cones. In order to define $u_1(x)u_2(x)$, it is enough to define it at each point of Ω. For $y \in \Omega$, there are convex cones G_j^ν $(\nu=1,2,\ j=1,\cdots,N_\nu)$ such that $\operatorname{supp} u_\nu \subset \bigcup_j G_j^\nu$, $G_1^1 = G_1^2 = \sqrt{-1}\, T^*M$ and that we have either $G_j^1 \cap G_k^2 \subset \sqrt{-1}\, T_M^*M$ or $G_j^{1a} \cap G_k^2 \not\ni y$. By Theorem 2.2.1 there are hyperfunctions v_j^ν such that $u_\nu = \sum_j v_j^\nu$ and $SS(v_j^\nu) \subset G_j^\nu$. We define $u_1u_2 = \sum sp(v_j^1 v_k^2)$ on a neighborhood of y, where the summation is taken over (j,k) satisfying $G_j^{1a} \cap G_k^2 \subset \sqrt{-1}\, T_M^*M$. Then this definition does not depend on the choice of G_j^ν and v_j^ν by the preceding theorem. We leave the details to the reader.

3.3. *Pull-back of microfunctions.* Let N and M be two real analytic manifolds and $f: N \to M$ a real analytic map. In this section we will describe how to pull back microfunctions from M to N via f. We shall denote by ρ the map $N \times_M \sqrt{-1}\,T^*M \to \sqrt{-1}\,T^*N$ and by $\tilde{\omega}$ the map $N \times_M \sqrt{-1}\,T^*M \to \sqrt{-1}\,T^*M$. We shall denote by $\sqrt{-1}\,T^*_N M$ the kernel of ρ.

THEOREM 3.3.1. (S.K.K. Th. 2.2.6, p. 292). *Let* u *be a hyperfunction on* M *such that* $\tilde{\omega}^{-1}SS(u) \cap \sqrt{-1}\,T^*_N M$ *is contained in the zero section* $N \times_M \sqrt{-1}\,T^*_M M$. *Then, the pull-back* $f^*(u)$ *of* u *is well-defined and we have* $SS(f^*(u)) \subset \rho\tilde{\omega}^{-1}SS(u)$.

THEOREM 3.3.2. (S.K.K. Th. 2.2.6). *Let* Ω_M *be an open set of* $\sqrt{-1}\,T^*M$, Ω_N *an open set of* $\sqrt{-1}\,T^*N$. *Let* u *be a microfunction defined on* Ω_M. *Suppose that* $\tilde{\omega}^{-1}(\mathrm{supp}\, u) \cap \rho^{-1}(\Omega_N) \to \Omega_N$ *is a proper map. Then, the pull-back* $f^*(u)$ *of* u *is well-defined as a microfunction on* Ω_N.

Proof. The proof is along the same lines as the proof of Theorem 3.2.1. In fact, the conditions on $SS(u)$ means that we can write u as a sum $\sum b_{\Omega_j}(\phi_j)$ so that $f^{-1}(\Omega_j)$ is large enough for the boundary value $b_{f^{-1}(\Omega_j)}(\phi_j \circ f)$ to be well defined. Thus, we can define $f^*(u) = \sum b_{f^{-1}(\Omega_j)}(\phi_j \circ f)$. The second theorem is also proved in the same way.

EXAMPLES. a) We can define the δ-function on $\mathbf{R}^n$ as the product $\delta(x) = \delta(x_1) \cdots \delta(x_n)$.

b) If $f: M \to \mathbf{R}$ satisfies $df \neq 0$ on $f^{-1}(0)$, then we can define

$$\delta(f(x)),\ f(x)_+^{\lambda},\ (f(x) \pm i0)^{\lambda},\ \cdots$$

as pull-backs of the hyperfunctions $\delta(t), t_+^{\lambda}, (t \pm i0)^{\lambda}$ of one variable by the map f. By definition, setting $\Omega_\pm = \{x \in X;\ \pm \mathrm{Im}\, f(x) > 0\}$, $(f(x) \pm i0)^{\lambda} = b_{\Omega_\pm}(f^{\lambda})$, $\delta(f(x)) = \frac{1}{-2\pi i}(b_{\Omega_+}(f^{-1}) - b_{\Omega_-}(f^{-1}))$.

c) Set $u = \frac{1}{4}(Y(x+y) + Y(x-y) - Y(-x+y) - Y(-x-y))$. (See §1.4 example c).) Then

$$\begin{cases} (D_x^2 - D_y^2)u = 0 \\ u|_{x=0} = 0 \\ \frac{\partial u}{\partial x}\Big|_{x=0} = \delta(y) . \end{cases}$$

(In this case,

$$SS(u) = \{(x, y; \xi, \eta); \begin{matrix} x^2 \geq y^2 \\ \xi^2 = \eta^2 \end{matrix}, (x-y)(\xi+\eta) = (x+y)(\xi-\eta) = 0\}$$

$$= \{(x, y; \xi, \eta); x^2 \geq y^2, \xi = \eta = 0\}$$

$$\cup \{(x, y; \xi, \eta); x = y, \xi = -\eta\}$$

$$\cup \{(x, y; \xi, \eta); x = -y, \xi = \eta\}$$

and $u|_{x=0}$ means the pull-back of u by the map $\mathbf{R} \to \mathbf{R}^2$ $(y \mapsto (0, y))$.

3.4. *Property of* $\partial/\partial t$. Consider $\mathbf{R}^{n+1} \ni (t, x) = (t, x_1, \cdots, x_n)$. Denote by j_a the injection $\mathbf{R}^n \to \mathbf{R}^{n+1}$ defined by $x \mapsto (a, x)$ and F the projection $\mathbf{R}^{n+1} \to \mathbf{R}^n$ given by $(t, x) \mapsto x$ $(a \in \mathbf{R})$. We shall investigate the properties of $\partial/\partial t$ as a micro-local operator.

PROPOSITION 3.4.1. i) $\partial/\partial t : \mathcal{C} \to \mathcal{C}$ *is surjective.*

ii) *if a microfunction* u *satisfies* $\partial/\partial t\, u = 0$ *at* $p = (t, x; i(\tau\, dt + \langle\xi, dx\rangle))$ *with* $\tau \neq 0$ *then* $u = 0$ *in a neighborhood of* p.

iii) *if a microfunction* u *satisfies* $\partial/\partial t\, u = 0$ *and is defined near* $(a, x; i(\langle\xi, dx\rangle))$, *then*

$$u = F^*(j_a^* u) .$$

Proof. In order to show $\partial/\partial t : \mathcal{C} \to \mathcal{C}$ is surjective, it is enough to show that $\partial/\partial t : \mathcal{B} \to \mathcal{B}$ is surjective. Since a hyperfunction is a sum of boundary values of holomorphic functions ϕ and since we may suppose that ϕ

is defined on a convex set Ω, $\partial/\partial t : \mathcal{B} \to \mathcal{B}$ is surjective, because $\partial/\partial t\, \mathcal{O}(\Omega) = \mathcal{O}(\Omega)$.

To prove ii) and iii) we must investigate more precisely the properties of $\partial/\partial t$ on domains in $\mathbf{C}^{n+1}$. Let Ω be a convex neighborhood of $(0,0)$. For a convex open cone V of $\mathbf{R}^{1+n}$, set $T_\Omega(V) = \{Z \in \Omega;\ \mathrm{Im}\, Z \in V\}$. Then we have

a) $\partial/\partial t\, \mathcal{O}(T_\Omega(V)) = \mathcal{O}(T_\Omega(V))$.

b) Suppose $u \in \mathcal{O}(T_\Omega(V))$ such that $\partial/\partial t\, u = 0$. Then

$$u \in \mathcal{O}(T_\Omega(F^{-1}FV)).$$

LEMMA 3.4.2. *Let* p *be a point in* $\pi^{-1}(0)$. *Then, for any hyperfunction* u *such that* $SSu \not\ni p$, *there is hyperfunction* v *such that* $\partial/\partial t\, v = u$ *and* $SSv \not\ni p$.

Proof. If $p \in \sqrt{-1}\, T^*_M M$, then $u = 0$ and hence the lemma is trivial. If $p \neq 0$, then setting $p = (0,0; \sqrt{-1}(\tau_0, \xi_0))$, we can write

$$u = \sum b_{V_j}(\phi_j)$$

with

$$\phi_j \in \mathcal{O}(T_\Omega(V_j))$$

with convex open cones V_j in $\mathbf{R}^{1+n}$ such that $\tau_0 t + \langle \xi_0, x\rangle > 0$ for any $(t,x) \in V_j$. We can solve $\partial/\partial t\, \psi_j = \phi_j$ with $\psi_j \in \mathcal{O}(T_\Omega(V_j))$. Then $v = \sum b_{V_j}(\psi_j)$ satisfies the desired condition.

Now let us prove (ii). Take a point $p = (0,0; i(\tau,\xi))$. Let u be a microfunction such that $\partial/\partial t\, u = 0$ at p. Take a hyperfunction $\tilde{u}$ such that $\mathrm{sp}\, \tilde{u} = u$ at p. Then $SS(\partial/\partial t\, \tilde{u}) \not\ni p$. Therefore, by Lemma 3.4.2, there is a hyperfunction, v such that $\partial/\partial t\, \tilde{u} = \partial/\partial t\, v$ and $SSv \not\ni p$. Thus, replacing $\tilde{u}$ with $\tilde{u} - v$, we can assume from the first that $\partial/\partial t\, \tilde{u} = 0$, and $u = \mathrm{sp}\, \tilde{u}$ at p. We shall write $\tilde{u} = \sum b_{V_j}(\phi_j)$ with

$\phi_j \in \mathcal{O}(T_\Omega(V_j))$. Since $\partial/\partial t\, \tilde{u} = 0$, there are convex open cones V_{jk} and $\phi_{jk} \in \mathcal{O}(T_\Omega(V_{jk}))$ such that $V_{jk} \supset V_j \cup V_k$, $V_{jk} = V_{kj}$, $\partial/\partial t\, \phi_j = \sum_k \phi_{jk}$ on $T_\Omega(V_j)$ and $\phi_{jk} = -\phi_{kj}$, by replacing V_j and Ω smaller ones if necessary. We can solve $\phi_{jk} = \partial\psi_{jk}/\partial t$ on $T_\Omega(V_{jk})$ such that $\psi_{jk} = -\psi_{kj}$. Hence $\partial/\partial t(\phi_j - \sum_k \psi_{jk}) = 0$. Thus replacing ϕ_j with $\phi_j - \sum_k \psi_{jk}$, we may assume from the first time $\partial/\partial t\, \phi_j = 0$. Hence ϕ_j is a function of x and hence defines an element of $\mathcal{O}(T_\Omega(F^{-1}FV_j))$. Thus $b(\phi_j) = F^*j_a{}^*(b(\phi_j))$ and $SS(b\phi_j) \not\ni (t, x; i(\tau, \xi))$ when $\tau \neq 0$. Therefore, the singular spectrum of $\tilde{u}$ does not contain such points, and $\tilde{u} = F^*j_a{}^*\tilde{u}$. This proves (ii) and (iii).

3.5. *Integration of microfunctions.* We shall next describe how to integrate microfunctions. Set $N = \mathbf{R}^{1+n}$, $M = \mathbf{R}^n$ and F, j_a as in §3.4. Then we have $\rho : N \underset{M}{\times} \sqrt{-1}\, T^*M \hookrightarrow \sqrt{-1}\, T^*N$ and $\tilde{\omega} : N \underset{M}{\times} \sqrt{-1}\, T^*M \to \sqrt{-1}\, T^*M$.

PROPOSITION 3.5.1. *Let Ω_M be an open set of $\sqrt{-1}\, T^*M$, Ω_N an open set of $\sqrt{-1}\, T^*N$, u a microfunction on N defined on Ω_N. Suppose that ρ^{-1} supp $u \cap \tilde{\omega}^{-1}(\Omega_M) \to \Omega_M$ is a proper map. Then the integral $F_*(u dt) = \int u(t, x)\, dt$ is well defined on Ω_M.*

We define

$$v(x) = \int u(t, x)\, dt$$

by an indefinite integral. Take a point $\gamma = (x_0, i\xi_0)$ of Ω_M. Then, there is a hyperfunction $\tilde{u}$ and $a < b$ such that

$$\text{sp}\, \tilde{u} = u \quad \text{on} \quad \tilde{\omega}^{-1}(\gamma) \cap p^{-1}\Omega_N\,,$$

$$SS\tilde{u} \cap \tilde{\omega}^{-1}(\gamma) \subset \text{supp}\, u\,,$$

$$SS\tilde{u} \cap \{(t, x_0;\ i(\tau, k\xi_0));\ k \geq 0,\ t \geq b \ \text{or}\ t \leq a,\ (\tau, k) \neq 0\} = \emptyset\,.$$

By Proposition 3.4.1, there is a hyperfunction w such that $\partial/\partial t\, w = \tilde{u}$. We define v at γ by

$$v = sp(j_b^* w - j_a^* w) .$$

It is easy to see that this definition does not depend on the choice of b, a and w.

We can also define integrations with respect to several variables as a succession of integrations with respect to one variable.

THEOREM 3.5.2. (S.K.K. Th. 2.3.1, p. 295). *Let* M *and* L *be two real analytic manifolds,* $N = M \times L$, F *the projection from* N *to* M *and* dt *a real analytic volume element on* L. *Let* Ω_M *and* Ω_N *be open sets of* $\sqrt{-1}\,T^*M$ *and* $\sqrt{-1}\,T^*N$, *respectively. Let* u *be a microfunction on* N *defined on* Ω_N. *Suppose that* $\rho^{-1}\operatorname{supp} u \cap \tilde{\omega}^{-1}\Omega_M \to \Omega_M$ *is a proper map. Then, the integral* $F_*(udt) = \int u(x,t)\,dt$ *is well defined as a microfunction on* Ω_M.

3.6. EXAMPLES. a) *The plane wave expansion of the* δ*-function*:

$$\delta(x) = \frac{(n-1)!}{(-2\pi\sqrt{-1})^n} \int_{S^{n-1}} \frac{\omega(\xi)}{(<x,\xi> + i0)^n}$$

where $\omega(\xi) = \xi_1 d\xi_2 \cdots d\xi_n - \xi_2 d\xi_1 d\xi_3 \cdots d\xi_n + \cdots + (-1)^{n-1}\xi_n d\xi_1 \cdots d\xi_{n-1}$, and $S^{n-1} = (\mathbf{R}^n - \{0\})/\mathbf{R}^+$ the $(n-1)$-dimensional sphere. This explains the formula

$$\delta(x) = \frac{1}{(-2\pi\sqrt{-1})^n} \sum b_{T(G_j)}(u_{G_j})$$

in Example b) §1.4.

b) $\int \delta(x)\,dx = 1$

because $\frac{dY(x)}{dx} = \delta(x)$ and $Y(x) = 1$ $(x>0)$, $Y(x) = 0$ $(x<0)$.

c) $SS\delta(x) = \{(x;\ i\xi);\ x=0\}$

because we know that $SS\delta(x) \subset \{(x;\ i\xi);\ x=0\}$.

Since for $g \in GL(n; \mathbf{R})$,

$$\delta(gx) = |\det g|^{-1}\delta(x),$$

$SS\delta(x)$ is invariant by the action of $GL(n; \mathbf{R})$. Suppose that $SS\delta(x) \not\ni (0; i(1,0\cdots 0))$. Then

$$\delta(x_1) = \int \delta(x)\,dx_2\cdots dx_n$$

is micro-analytic at $(0, \sqrt{-1}\,dx_1)$, which is a contradiction. Hence $SS(\delta) \ni (0; i(1,0\cdots 0))$ and hence any $(0; i\xi)$ $(\xi \neq 0)$.

d) $\displaystyle\int \frac{1}{t - x^2 + i0}\,dx = -\pi i(t+i0)^{-1/2}$. Note that this integral has a sense at $(0, \sqrt{-1}\,dt)$. For example, $u(t,x) = (t-x^2)^{-1}$ is holomorphic on

$$\Omega = \{(t,x) \in \mathbf{C}^2;\ \operatorname{Im} t > 2|\operatorname{Im} x|,\ |\operatorname{Re} x| < 1\}.$$

Fix $0 < a < 1$. Set

$$v(t,x) = \int_0^x u(t,x)\,dx.$$

Then v is also defined on Ω and $\partial v/\partial x = u$. Therefore, we have

$$F = \int \frac{dx}{t - x^2 + i0} = v(t+i0, a) - v(t+i0, -a).$$

For $\operatorname{Im} t > 0$, $v(t,a) - v(t,-a) = \int_{-a}^{a} u(t,x)\,dx$. Let γ_+ be the path $ae^{i\theta}$ $(0 \le \theta \le \pi)$. (See the figure below.) Then $\int_{\gamma_+} v(t,x)\,dx$ is holomorphic at $t = 0$. Set $w(t) = \oint u(t,x)\,dx$ for $\operatorname{Im} t > 0$ where $\oint$ is the contour integral around $\sqrt{t}$. Then

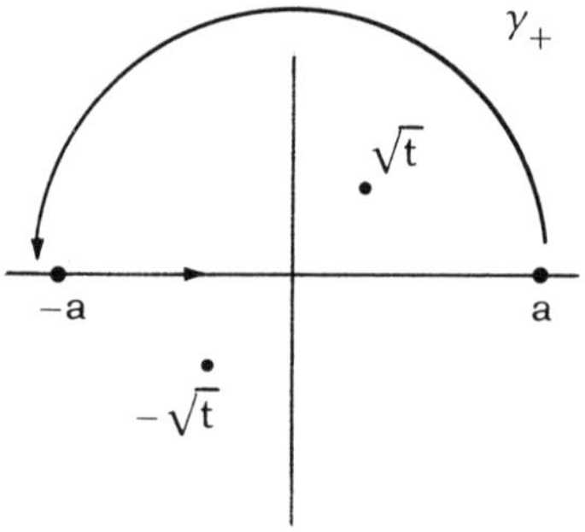

x-plane

$$\int \frac{dx}{t - x^2 + i0} = w(t+i0) \quad \text{at} \quad (0, \sqrt{-1}\, dt)\ ,$$

$$w(t) = -\frac{\pi i}{\sqrt{t}}\ ,$$

and hence we have

$$\int \frac{1}{t - x^2 + i0}\, dx = -\pi i(t+i0)^{-1/2} \quad \text{at} \quad (0, \sqrt{-1}\, dt)\ .$$

Changing i to $-i$ we obtain

$$\int \frac{dx}{t - x^2 - i0} = \pi i(t-i0)^{-1/2} \quad \text{at} \quad (0, -\sqrt{-1}\, dt)\ .$$

Changing t to $-t$, we obtain

$$\int \frac{dx}{t + x^2 \pm i0} = \pi(t \pm i0)^{-1/2} \quad \text{at} \quad (0, \pm\sqrt{-1}\, dt)\ .$$

e) $$\int (t-x+i0)^{\lambda-1}(x+i0)^{\mu-1}\, dx = (-2\pi i)\,\frac{\Gamma(1-\lambda-\mu)}{\Gamma(1-\lambda)\Gamma(1-\mu)}\,(t+i0)^{\lambda+\mu-1}\ .$$

First, this formula has sense at $(0, idt)$. Assume first $\mathrm{Re}(\lambda+\mu) < 1$. The integral is by definition equal to

$$\mathrm{sp}\left(b\int_{-a}^{a}(t-x)^{\lambda-1}x^{\mu-1}dx\right)$$

over the contour indicated in the figure below.

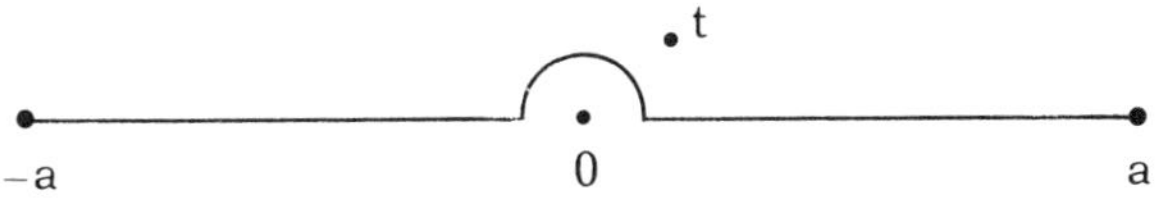

Here $\int_{-a}^{a}(t-x)^{\lambda-1}x^{\mu-1}dx$ is holomorphic on $\{t\in\mathbf{C};\ \mathrm{Im}\ t>0\}$. If $\mathrm{Re}(\lambda+\mu) < 0$, then $\int_{-\infty}^{-a}(t-x)^{\lambda-1}x^{\mu-1}dx$ and $\int_{a}^{\infty}(t-x)^{\lambda-1}x^{\mu-1}dx$ are holomorphic on t at the origin. Therefore, the integral is equal to

$$\int_{-\infty}^{\infty}(t-x)^{\lambda-1}x^{\mu-1}dx = -2\pi i\,\frac{\Gamma(1-\lambda-\mu)}{\Gamma(1-\lambda)\Gamma(1-\mu)}\,(t+i0)^{\lambda+\mu-1}\ .$$

By the analytic continuation on λ and μ we get the desired result.

f) $$\int_{\mathbf{R}^n}(t+\langle Ax, x\rangle+i0)^{\lambda}dx = \frac{e^{-(\pi i/2)q}}{|\det A|^{1/2}}\,\pi^{n/2}\,\frac{\Gamma(-\lambda-n/2)}{\Gamma(-\lambda)}\,(t+i0)^{\lambda+n/2}$$

at $(0, idt)$, where A is a non-degenerate symmetric matrix and q is the number of negative eigenvalues of A. By a coordinate transform, we may assume that A is a diagonal matrix. Then, by the succession of integrals, it is enough to show

$$\int(t+ax^2+i0)^{\lambda}dx = \frac{\varepsilon}{|a|^{1/2}}\,\frac{\sqrt{\pi}\,\Gamma(-\lambda-1/2)}{\Gamma(-\lambda)}\,(t+i0)^{\lambda+1/2}$$

$\varepsilon = 1\ (a>0)$, $\varepsilon = -i\ (a<0)$.

Changing $x \mapsto |a|^{-1/2}x$, we may assume $a = \pm 1$. Then we know already

$$\int \frac{dx}{t+ax^2+i0} = \varepsilon\pi(t+i0)^{-1/2}.$$

Therefore, we have

$$\int (t+ax^2+i0)^{\lambda}dx = \frac{1}{-2\pi i}\iint (t-s+i0)^{\lambda}(s+ax^2+i0)^{-1}ds\,dx$$

$$= \frac{\varepsilon\pi}{-2\pi i}\int (t-s+i0)^{\lambda}(s+i0)^{-1/2}ds$$

$$= \varepsilon\sqrt{\pi}\,\frac{\Gamma(-\lambda-1/2)}{\Gamma(-\lambda)}(t+i0)^{\lambda+1/2}.$$

§4. *Micro-differential operators*

4.1. *Micro-local operators.* Let M and N be manifolds. Fix a real analytic density dy on N. We denote by p_1 the projection $\sqrt{-1}\,T^*(M\times N) \to \sqrt{-1}\,T^*M$ (resp. p_2^a the projection $\sqrt{-1}\,T^*(M\times N) \to \sqrt{-1}\,T^*N$) defined by $(x, y; i(\xi,\eta)) \mapsto (x, i\xi)$ (resp. $(y, -i\eta)$).

Let Ω be an open set in $\sqrt{-1}\,T^*(M\times N)$, Z a closed set of Ω.

PROPOSITION 4.1.1. *Let* Ω_N *be an open set of* $\sqrt{-1}\,T^*N$, Ω_M *an open set of* $\sqrt{-1}\,T^*M$. *Suppose that*

$$p_1^{-1}\Omega_M \cap (p_2^a)^{-1}\Omega_N \cap Z \to \Omega_N$$

is a proper map. Then for a microfunction $K(x,y)$ *defined on* Ω *and a microfunction* $v(y)$ *defined on* Ω_N, $u(x) = \int K(x,y)v(y)dy$ *is well defined on* Ω_M *if* $\operatorname{supp} K \subset Z$.

This follows easily from Theorems 3.2.2, 3.3.2, and 3.5.2.

COROLLARY 4.1.2. *Suppose that*

$$p_1^{-1}(\Omega_M) \cap Z \cap (p_2^a)^{-1}(\Omega_N) \to \Omega_N \tag{4.1}$$

and

$$p_2^{-1}(\Omega_M) \cap Z \cap (p_2^a)^{-1}(\Omega_N) \to \Omega_M$$

are homeomorphisms. Let Φ *be the map* $\Omega_M \to \Omega_N$ *defined by these two homeomorphisms. Then, for any microfunction* $v(y)$ *defined on an open subset* U *of* Ω_N, $u(x) = \int K(x,y)\,v(y)\,dy$ *is well defined on* $\Phi^{-1}(U)$, *i.e. this defines a sheaf homomorphism* $\Phi^{-1}(\mathcal{C}_N|_{\Omega_N}) \to \mathcal{C}_M$.

Consider the particular case $N = M$. Let Δ be the diagonal in $M \times M$. Set $Z = \sqrt{-1}\,T^*_\Delta(M \times M) = \{(x, x, \sqrt{-1}(\xi, -\xi))\}$. We identify Z with $\sqrt{-1}\,T^*M$ by p_1. Let Ω be an open set of $\sqrt{-1}\,T^*M$. If $K(x,y)$ is a microfunction defined on $\Omega \times \Omega$ whose support is contained in Z, then, for any microfunction $v(x)$ defined on an open subset U of Ω, $\int K(x,y)\,v(y)\,dy$ is well defined on U. Therefore

$K: v(x) \mapsto \displaystyle\int K(x,y)\,v(y)\,dy$ gives a sheaf homomorphism $\mathcal{C}|_\Omega \to \mathcal{C}|_\Omega$. We call K a *micro-local operator* on Ω and $K(x,y)$ the kernel function of K.

The identity operator is a micro-local operator corresponding to the kernel function $\delta(x-y) = \prod_{j=1}^{n} \delta(x_j - y_j)$. A differential operator $P(x,D)$ is a micro-local operator corresponding to the kernel function $P(x,D)\delta(x-y)$.

4.2. *Micro-differential operators (real case).* The class of micro-local operators is too wide a class to work with effectively. We shall introduce a class of micro-local operators, called micro-differential operators, which we can manipulate easily. This class is in some sense, a localization of differential operators.

First, we shall investigate the kernel functions of differential operators.

Let $P(x,D) = \sum_{|\alpha| \le m} a_\alpha(x) D^\alpha$, $D^\alpha = \dfrac{\partial^{|\alpha|}}{\partial x_1^{\alpha_1} \cdots \partial x_n^{\alpha_n}}$ be a differential operator with real analytic coefficient on $\mathbf{R}^n$. Set $P_j(x,\xi) = \sum_{|\alpha|=j} a_\alpha(x)\xi^\alpha$.

THEOREM 4.2.1. *$P : \mathcal{C} \to \mathcal{C}$ is bijective on*

$$\Omega = \{(x; i\xi) \in \sqrt{-1}\, T^*M;\ P_m(x, i\xi) \neq 0\}\ .$$

The proof of this involves constructing the inverse of P on Ω explicitly. This turns out to be a micro-differential operator.

Let us put

$$\Phi_\lambda(z) = \frac{\Gamma(\lambda)}{(-z)^\lambda}\ ,$$

where we take its branch on $z \in \mathbf{C} - \{z;\ z \ge 0\}$ such that $\Phi_\lambda(-1) = \Gamma(\lambda)$.

When $\lambda = -m\ (m = 0, 1, 2, \cdots)$, we set $\Phi_\lambda(z) = -\frac{1}{m!} z^m \left(\log(-z) - \sum_{j=1}^m \frac{1}{j} + \gamma\right)$ where γ is the Euler constant. Then $(\partial/\partial z)\Phi_\lambda(z) = \Phi_{\lambda+1}(z)$. Now, consider the kernel function of P, that is, $P(x,D)\delta(x-y)$. We have

$$\delta(x-y) = \frac{(n-1)!}{(-2\pi i)^n} \int \frac{\omega(\xi)}{(<x-y, \xi> + i0)^n}$$

$$= \frac{1}{(2\pi)^n} \int \Phi_n(<x-y, i\xi> - 0)\,\omega(\xi)\ .$$

Here $\Phi_n(<x, i\xi> - 0)$ is a boundary value of holomorphic function

$$\Phi_n(<x, i\xi>)$$

from the direction $\mathrm{Re} <x, i\xi> < 0$ (i.e. $\mathrm{Im} <x, \xi> > 0$). Then we have

$$P(x, D_x)\delta(x-y) = \frac{1}{(2\pi)^n}\int \sum_{j\geq 0} P_j(x, i\xi)\Phi_{n+j}(<x, i\xi> - 0)\omega(\xi) .$$

This formula suggests introducing the following class of operators (called micro-differential operators): Let $(x_0, i\xi_0)$ be a point of $\sqrt{-1}\ T^*M \subset T^*X$ ($M \subset \mathbf{R}^n$, $X \subset \mathbf{C}^n$). Let $\lambda \in \mathbf{C}$.

Let $\{P_{\lambda+j}(z, \zeta)\}_{j\in\mathbf{Z}}$ be a series of holomorphic functions defined in a neighborhood U of $(x_0, i\xi_0)$ in $\mathbf{C}^n \times \mathbf{C}^n$. Suppose that $P_{\lambda+j}(z, \zeta)$ is homogeneous of degree $(\lambda + j)$ with respect to ζ. Consider the kernel function

$$K(x, y) = \frac{1}{(2\pi)^n}\int \sum_j P_{\lambda+j}(x, i\xi)\Phi_{n+\lambda+j}(<x-y, i\xi> - 0)\omega(\xi) .$$

To make sense of the integrand set

$$K(x, y, \zeta) = \sum_j P_{\lambda+j}(x, \zeta)\Phi_{n+\lambda+j}(<x-y, \zeta>)$$

for $\mathrm{Re} <x-y, \zeta> < 0$. We impose conditions so that this series converges on

$$\Omega_\varepsilon : \{(x, y, \zeta);\ (x, \zeta) \in U;\ |y-x_0| << 1 ,$$
$$\mathrm{Re} <x-y, \zeta> > -\varepsilon|\mathrm{Im} <x-y, \zeta>|\} ;$$

i.e. we assume for $\forall\varepsilon$, there is a C_ε such that

$$(4.2.1) \qquad |P_{\lambda+j}(z, \zeta)| \leq \frac{C_\varepsilon}{j!}\varepsilon^j \quad \text{for} \quad \forall j > 0 ,$$

and a constant R such that

$$|P_{\lambda+j}(z, \zeta)| \leq (-j)!R^{-j} \quad \text{for } \forall j < 0 .$$

Since $K(x, y, \zeta)$ is essentially a Laurent series in $<x-y, \zeta>$, this growth condition assures the convergence of $K(x, y, \zeta)$. Thus, we can define a hyperfunction $K(x, y, i\xi) = b_{\Omega_\varepsilon}(K(x, y, i\xi))$ the boundary value being with respect to $\{(x, y, \xi);\ \mathrm{Re} < x-y, i\xi> < 0\}$. Therefore, we have

$$SSK \subset \{(x,y,\xi;\ i(\tilde{\xi},\eta,\rho));\ k<x-y,\xi> = 0,\ \eta = -\tilde{\xi},\ \rho = k(x-y),\ \tilde{\xi} = k\xi \text{ for some } k \geq 0\}.$$

Hence

$$K(x, y) = \frac{1}{(2\pi)^n} \int K(x, y, i\xi)\omega(\xi)$$

is well defined in a neighborhood of $(x_0, x_0;\ i(\xi_0, -\xi_0))$ as a microfunction and the support is contained in $\sqrt{-1}\, T^*_M(M \times M) = \{(x, y;\ i(\xi, \eta));\ x = y, \xi = -\eta\}$. This shows that K is the kernel of a micro-local operator.

We call the micro-local operator K a *micro-differential operator* and denote it by $\sum_{j \in \mathbf{Z}} P_{\lambda+j}(x, D_x)$. Differential operators are special examples of micro-differential operators.

THEOREM 4.2.2. *Let* $P(x, D) = \sum P_{\lambda+j}(x, D)$ *and* $Q = \sum Q_{\mu+j}(x, D)$ *be micro-differential operators. Then, the composition* $R = PQ$ *is also a micro-differential operator and*

$$R = \sum R_{\lambda+\mu+j}(x, D)$$

$$R_{\lambda+\mu+\ell}(z, \zeta) = \sum_{\ell = j+k-|a|} \frac{1}{a!} (D^a_\zeta P_{\lambda+j})(D^a_z Q_{\mu+k})$$

$(D^a_\zeta = \partial^{|a|}/\partial\zeta_1^{a_1} \cdots \partial\zeta_n^{a_n},\ a! = a_1! \cdots a_n!)$.

THEOREM 4.2.3. *Let* $P = \sum P_{\lambda+j}(x, D)$ *be a micro-differential operator with the kernel* $K(x, y)$. *Then the formal adjoint operator* ${}^tP$ *(i.e. operator whose kernel is* $K(y, x)$*) is also a micro-differential operator and we have* ${}^tP = \sum Q_{\lambda+j}(x, D)$ *with*

$$Q_{\lambda+j}(z,\zeta) = \sum_{j=k-|\alpha|} \frac{(-1)^{|\alpha|}}{\alpha!} (D_z^\alpha D_\zeta^\alpha P_{\lambda+k})(z,-\zeta) .$$

4.3. *Micro-differential operators in a complex domain.* Let X be a complex manifold. We shall define micro-differential operators on an open set Ω of T^*X as follows: a micro-differential operator P on Ω is a series $\{P_j(z,\zeta)\}_{j\in Z}$ such that

(4.3.1) $$P_j(z,\zeta)$$

is a holomorphic function defined on Ω, homogeneous of degree j with respect to ζ and

(4.3.2) $$\{P_j(z,\zeta)\}$$

satisfies the following growth conditions:

(4.3.2.1) for every $\varepsilon > 0$ and every compact set $K \subset \Omega$, there is a constant $C_{K,\varepsilon} > 0$ such that

$$|P_j(z,\zeta)| \le \frac{C_{K,\varepsilon}}{j!} \varepsilon^j \quad \text{for} \quad j > 0,\ (z,\zeta) \in K .$$

(4.3.2.2) for every compact set $K \subset \Omega$, there is a constant $R_K > 0$ such that

$$|P_j(z,\zeta)| \le (-j)!\, R_K^{-j} \quad \text{for} \quad j < 0,\ (z,\zeta) \in K .$$

We shall write $\sum P_j(z,D)$ for $\{P_j(z,\zeta)\}$.

We let $\mathcal{E}_X^\infty(\Omega)$ be the set of micro-differential operators on Ω. Then $\Omega \mapsto \mathcal{E}_X^\infty(\Omega)$ is a sheaf on T^*X. If X is the complexification of a real analytic manifold M, then $P \in \mathcal{E}_X^\infty(\Omega)$ operates on $\mathcal{C}_M(\sqrt{-1}\, T^*M \cap \Omega)$. (This is just like the relation between differential operators with real analytic coefficients and those with holomorphic coefficients.) We denote by $\mathcal{E}_X(m)$ the subsheaf of $\mathcal{E}_X^\infty$ consisting of $P = \sum P_j(z,D)$ such that $P_j = 0$ for $j > m$, and let $\mathcal{E}_X = \cup\, \mathcal{E}_X(m)$. We shall denote by $\mathcal{O}_{T^*X}(m)$ the sheaf of holomorphic functions homogeneous of degree m with respect to ζ and by σ_m the homomorphism $\mathcal{E}_X(m) \to \mathcal{O}_{T^*X}(m)$ which sends P to P_m. We have $\mathcal{E}_X(m)/\mathcal{E}_X(m-1) \xrightarrow{\sim} \mathcal{O}_{T^*X}(m)$. We define the product of two sections of $\mathcal{E}^\infty$ by the formula in Theorem 4.2.2. This gives the structure of a ring to $\mathcal{E}_X^\infty$, and $\mathcal{E}_X$ becomes a sub-ring. Moreover, we have

(4.3.3) $$\mathcal{E}_X(m_1)\mathcal{E}_X(m_2) \subset \mathcal{E}_X(m_1+m_2)$$

$$\text{and } \sigma_{m_1+m_2}(PQ) = \sigma_{m_1}(P)\sigma_{m_2}(Q)$$

$$\text{for } P \in \mathcal{E}_X(m_1) \text{ and } Q \in \mathcal{E}_X(m_2).$$

(4.3.4) For $P \in \mathcal{E}_X(m_1)$ and $Q \in \mathcal{E}_X(m_2)$, $[P, Q] = PQ - QP$ belongs to $\mathcal{E}_X(m_1+m_2-1)$ and $\sigma_{m_1+m_2-1}([P, Q]) = \{\sigma_{m_1}(P), \sigma_{m_2}(Q)\}$, where

$$\{f, g\}(z, \zeta) = \sum_j \left(\frac{\partial f}{\partial \zeta_j}\frac{\partial g}{\partial z_j} - \frac{\partial g}{\partial \zeta_j}\frac{\partial f}{\partial z_j}\right).$$

4.4. *Properties of micro-differential operators.* The ring of micro-differential operators is in some sense the localization of the ring of differential operators (just as the ring of holomorphic functions is a localization of the ring of polynomials). For example, we have

THEOREM 4.4.1. (S.K.K. Thm. 2.1.1, p. 356). *Let* $P(x, D)$ *be a micro-differential operator of order* $\leq m$ *and suppose that* $\sigma_m(P)$ *does not vanish at a point* p *in* T^*X. *Then, there is an inverse* P^{-1} *in* $\mathcal{E}(-m)$ (*i.e.* $P^{-1}P = PP^{-1} = 1$) *in a neighborhood of* p.

This implies immediately Theorem 4.2.1.

§5. *Quantized contact transforms*

5.1. *Complex case.* (S.K.K. Chap. II, §4.3). Consider the ring of micro-differential operators on $\mathbf{C}^n$ defined in a neighborhood of $(0, dx_1)$. Any element P of this ring can be written in the form

$$\sum_{\alpha,\beta} a_{\alpha,\beta} x^\alpha D^\beta$$

where $\alpha = (\alpha_1, \cdots, \alpha_n)$, $\beta = (\beta_1, \cdots, \beta_n)$ with $\alpha_j \geq 0$ $(j = 1, \cdots, n)$, $\beta_j \geq 0$ $(j = 2, \cdots, n)$. Therefore, this ring is a kind of completion (or localization) of the ring generated by x_j and D_j. Note that x_j and D_j satisfy the commutation relation

(5.1.1)
$$[x_j, x_k] = 0$$
$$[D_j, x_k] = \delta_{jk}$$
$$[D_j, D_k] = 0\,.$$

Let $\{P_j, Q_j\}_{j=1,\cdots,n}$ be 2n micro-differential operators defined in a neighborhood of (x_0, ξ_0) and satisfying the relations

(5.1.2)
$$[Q_j, Q_k] = [P_j, P_k] = 0$$
$$[P_j, Q_k] = \delta_{jk}\,.$$

We want to construct a ring homomorphism

$$\Psi : \mathcal{E} \to \mathcal{E}$$

mapping

$$\sum a_{\alpha,\beta} x^\alpha D^\beta \quad \text{onto} \quad \sum a_{\alpha,\beta} P^\alpha Q^\beta\,.$$

Note that condition (5.1.2) implies

(5.1.3)
$$\{\sigma(Q_j), \sigma(Q_k)\} = \{\sigma(P_j), \sigma(P_k)\} = 0\,,$$
$$\{\sigma(P_j), \sigma(Q_k)\} = \delta_{jk}\,.$$

Therefore $(x, \xi) \mapsto (\sigma(P_1), \cdots, \sigma(P_n), \sigma(Q_1), \cdots, \sigma(Q_n))$ defines a symplectic transformation.

THEOREM 5.1.1. *Let* X *be an* n-*dimensional complex manifold and* Ω *an open set of* T^*X. *Let* $\{P_j, Q_j\}_{1 \leq j \leq n}$ *be micro-differential operators. Assume that the condition* (5.1.2) *holds and that* P_j's *are of order* 1 *and* Q_j's *are of order* 0. *Let* Φ *be the symplectic transformation from* Ω *to* $T^*\mathbf{C}^n$ *defined by* $(x; \xi) \to (\sigma(Q_1), \cdots, \sigma(Q_n); \sigma(P_1), \cdots, \sigma(P_n))$. *Then there is a unique* $\mathbf{C}$-*Algebra homomorphism*

$$\Psi : \Phi^{-1} \mathcal{E}_{\mathbf{C}^n} \to \mathcal{E}_X|_\Omega$$

such that $\Psi(x_j) = Q_j$ *and* $\Psi(D_j) = P_j$. *Moreover,* Ψ *is necessarily an isomorphism.*

Note that Φ is a transform which preserves the canonical 1-form on the cotangent bundles. We call $(\Phi; \Psi)$ a quantized contact transformation. We have the relations

$$\Psi \mathcal{E}_{\mathbf{C}^n}(m) = \mathcal{E}_X(m) \tag{5.1.4}$$

(i.e. Ψ preserves order) and

$$\text{for } R \in \mathcal{E}_{\mathbf{C}^n}(m),\ \sigma_m(\Psi(R)) = \sigma_m(R) \circ \Phi \tag{5.1.5}$$

(i.e. Ψ preserves principal symbols).

THEOREM 5.1.2. *Let* Φ *be a homogeneous symplectic transformation from an open set* Ω *in* T^*X *into* $T^*\mathbf{C}^n$ (*i.e.* Φ *preserves the canonical* 1-*form*), *and* p *a point of* Ω. *Then, in a neighborhood of* p, *we can find a quantized contact transformation* $(\Phi; \Psi)$. (*Note that this* Ψ *is not unique.*)

This theorem decomposes the study of system of micro-differential operators into two parts; geometric and analytic. Let us consider a system of micro-differential equations

$$(\#) \qquad P_1 u = \cdots = P_N u = 0 .$$

Let $\mathcal{J}$ be the left Ideal of $\mathcal{E}$ generated by $P_1, \cdots, P_N$; we call characteristic variety V of (#) the analytic subset defined by $\{p \in T^*X; \sigma(R)(p) = 0 \text{ for } \forall R \in \mathcal{J}\}$. The characteristic variety cannot be an arbitrary subset of T^*X, in fact it has the following very special property:

THEOREM 5.1.3. (S.K.K. Thm. 5.3.2, p. 453). *The characteristic variety* V *of a system of micro-differential equation is involutive* (*i.e. if two holomorphic functions* f *and* g *vanish on* V, *then their Poisson bracket* $\{f, g\}$ *vanishes on* V).

If all holomorphic functions f which vanish on V are written in the form $\sigma(R)$ for $R \in \mathcal{J}$, then this theorem is obvious because $\{\sigma(P), \sigma(Q)\} = \sigma([P, Q])$ and $[P, Q] \in \mathcal{J}$ if $P, Q \in \mathcal{J}$. If it is not a case, the proof is complicated. At a "generic" point of V, we can transform V by a homogeneous symplectic transform into the subset

$$V_0 = \{(x, \xi); \xi_1 = \cdots = \xi_\ell = 0\} .$$

Therefore, by virtue of Theorem 5.1.2, it is enough to investigate systems of micro-differential equations whose characteristic variety is V_0.

By S.K.K. Thm. 5.3.7, p. 455, we know that these types of systems are of the form $D_1 u = \cdots = D_\ell u = 0$ (essentially) at a generic point, up to multiplication by operators in $\mathcal{E}^\infty$. Thus, we can say that at a generic point of its characteristic variety, every system is equivalent to the partial de Rham system, $D_1 u = \cdots = D_\ell u = 0$.

5.2. *Quantized contact transforms (real case).* If we are interested in the properties of microfunction solutions of a system of micro-differential equations, the results described in the last section require further elaboration.

Let M be a real analytic manifold of dimension n, X it complexification and $(\Phi;\Psi)$ a quantized contact transformation from an open set Ω of T^*X into $T^*\mathbf{C}^n$. Suppose that

$$\Phi(\Omega \cap \sqrt{-1}\,T^*M) \subset \sqrt{-1}\,T^*\mathbf{R}^n\,, \tag{5.2.1}$$

and set

$$P_j = \Psi(D_j)\,, \quad Q_j = \Psi(x_j)\,.$$

We shall consider the system of equations

$$\begin{aligned} x_j K(y,x) &= Q_j K(y,x) \qquad (x \in \mathbf{R}^n,\ y \in M) \\ -\frac{\partial}{\partial x_j} K(y,x) &= P_j K(y,x)\,. \end{aligned} \tag{5.2.2}$$

The characteristic variety of this system is

$$\Lambda = \{(p, \Phi(p)^a) \in T^*(X \times \mathbf{C}^n);\ p \in \Omega\}\,,$$

where $(x,\xi)^a = (x,-\xi)$. Therefore, if there is a microfunction solution K of (5.2.2), the support of K is contained in Λ. By Corollary 4.1.2,

$$K:\ v(x) \mapsto \int K(y,x)\,v(x)\,dx$$

gives rise to a sheaf homomorphism

$$K: \Phi^{-1}\mathcal{C}_{\mathbf{R}^n} \to \mathcal{C}_M|_\Omega\,.$$

Moreover, we have

$$K(x_j v)(y) = \int K(y,x)\,x_j\,v(x)\,dx$$

$$= \int Q_j K(y,x)\,v(x)\,dx$$

$$= Q_j(Kv)$$

and

$$K\frac{\partial}{\partial x_j}v = \int K(y,x)\left(\frac{\partial v}{\partial x_j}\right)dx$$

$$= -\int\left(\frac{\partial}{\partial x_j}K(y,x)\right)v(x)\,dx$$

$$= \int (P_j K(y,x))\,v(x)\,dx = P_j(Kv)$$

(by interation by parts). In fact, we have

THEOREM 5.2.1. (S.K.K. Chap. III, §1.3). *Let* p *be a point in* $\Omega \cap \sqrt{-1}\,T^*M$.

a) *Then, in a neighborhood of* $(p, \Phi(p)^a)$ *there is a microfunction solution* $K(y,x)$ *of* (5.2.2), *unique up to constant multiple.*

b) *If* K *is a non-zero solution of* (5.2.2), *the corresponding homomorphism* $K: \Phi^{-1}\mathcal{C}_{R^n} \to \mathcal{C}_M|_\Omega$ *is an isomorphism and we have*

$$K(Ru) = \Psi(R)(Ku)$$

for $u \in \mathcal{C}_{R^n}$ *and* $R \in \mathcal{E}^\infty$.

By this theorem, if we can find a *real* homogeneous symplectic transform which transforms the characteristic variety of a system of equations into a simple form, then it is enough to investigate equations whose characteristic variety has that simple form.

EXAMPLE 5.2.2. Consider a single equation $Pu = 0$. Suppose that $\sigma(P)$ is real and $d\sigma(P)$ is not parallel to $\omega = \sum \zeta_j dz_j$. We may suppose from the first that P is of order 1. Then by a *real* homogeneous contact transform we can transform $\sigma(P)$ to ξ_1. Thus, we may assume $\sigma(P) = \sigma(D_1)$. We can prove also (S.K.K. Thm. 2.1.2, p. 359) that there is an invertible micro-differential operator R such that $P = RD_1R^{-1}$. Hence, $Pu = 0$ is equivalent to $D_1 u = 0$. Thus, for example, we can conclude that

a) $Pu = v$ is locally solvable.

b) The support of a microfunction solution of $Pu = 0$ is a union of bicharacteristic strips of P.

EXAMPLE 5.2.3. (S.K.K. Chap. III, §2.3). Let $Pu = 0$ be a single equation. Suppose that $\{\sigma(P), \overline{\sigma(P)}\} \neq 0$. Then, by a *real* homogeneous contact transform, we can transform $\sigma(P)$ to $\xi_1 \pm i x_1 \xi_n$. Thus, in order to study the properties of $Pu = 0$, it is enough to investigate the property of the equation $(D_1 \pm \sqrt{-1}\, x_1 D_n) u = 0$. For example, we have

a) If $\{\sigma(P), \overline{\sigma(P)}\} > 0$, then $P : \mathcal{C} \to \mathcal{C}$ is surjective.

b) If $\{\sigma(P), \overline{\sigma(P)}\} < 0$, then $P : \mathcal{C} \to \mathcal{C}$ is injective.

Here, $\overline{\sigma(P)}$ is the complex conjugate of holomorphic function $\sigma(P)$ with respect to the real form $\sqrt{-1}\, T^*M$ of T^*X.

SOME APPLICATIONS OF BOUNDARY VALUE PROBLEMS FOR ELLIPTIC SYSTEMS OF LINEAR DIFFERENTIAL EQUATIONS

Masaki Kashiwara* and Takahiro Kawai*

§0. The purpose of this report is to show some applications of the theory of boundary value problems for elliptic systems of linear differential equations formulated in Kashiwara-Kawai [1]. The materials discussed in this report are contained in [1], [3], [5] and [6].

Hyperfunctions and microfunctions are "boundary value of holomorphic functions," while holomorphic functions are nothing but the solutions of Cauchy-Riemann equations, a typical example of elliptic systems. On the other hand, one of the basic results proved in S-K-K [8] asserts that any overdetermined system of micro-differential equations can be brought micro-locally to the canonical form

$$\mathfrak{M}_p : \begin{cases} \left(\dfrac{\partial}{\partial x_j} + \sqrt{-1}\,x_j \dfrac{\partial}{\partial x_n}\right)u = 0\,, & j = 1, \cdots, p \\[2ex] \left(\dfrac{\partial}{\partial x_j} - \sqrt{-1}\,x_j \dfrac{\partial}{\partial x_n}\right)u = 0\,, & j = p+1, \cdots, d \end{cases}$$

*Supported in part by NSF grant MCS77-18723.

Seminar on Micro-Local Analysis
0-691-08228-6/79/00 0039-23 $01.15/1 (cloth)
0-691-08232-4/79/00 0039-23 $01.15/1 (paperback)
For copying information, see copyright page

under moderate conditions on the system, while system $\mathfrak{M}_p$ is the tangential system of apparently simple system

$$\mathfrak{N}: \begin{cases} \dfrac{\partial}{\partial x_j} u = 0, & j = 1, \cdots, d(\leq n-1) \\ \left(\dfrac{\partial}{\partial x_n} + \sqrt{-1}\,\dfrac{\partial}{\partial x_{n+1}}\right) u = 0 & \end{cases}$$

induced on $S_p = \left\{ x \in R^{n+1};\ x_{n+1} + \frac{1}{2}\left(\sum_{j=1}^{p} x_j^2 - \sum_{j=p+1}^{d} x_j^2\right) = 0 \right\}$.

These observations naturally lead us to the expectation that the boundary value problems for (elliptic) systems will be most neatly formulated in the framework of hyperfunctions and microfunctions and that such a formulation will provide us with an effective tool for the study of the structure of solutions of general systems (appearing as tangential systems). In fact, such an expectation was realized in [1]-[6].

The discussion given in this report seems to be closely related in its spirit to the theory and applications of Töplitz operators done by Boutet de Monvel and Guillemin.

In this report we use the same notations as in [3] and S-K-K [8] with the exception that the sheaf of the micro-differential operators of finite order is denoted by $\mathcal{E}$ instead of $\mathcal{P}^f$.

§1. First let us recall the formulation of the problem given in [1]. Even though higher codimensional boundary is considered there, we restrict ourselves here to the case where the boundary is of codimension 1. This makes the presentation much simpler. Note also that the proof of the theorem becomes fairly easy in this case, even though we do not give it here. (See [3] §1.)

The situation which we discuss is the following:

Let M be a real analytic manifold and N its submanifold of codimension 1. Since the problem shall be formulated locally, we assume $M = \mathbf{R}^n$ and $N = \{x = (x_1, x') \in \mathbf{R}^n;\ x_1 = 0\}$. Let X (resp. Y) be a complexification of M (resp. N). We define $M_\pm$ by $\{x \in M;\ \pm x_1 > 0\}$ and $Z_\pm$ by $M_\pm \sqcup N$.

Let $\mathfrak{M}$ be an elliptic system of linear differential equations defined on M whose domain of definition extends over X. Hence it follows from the definition that

(1) $$SS(\mathfrak{M}) \cap \sqrt{-1}\,S^*M = \emptyset\ .^{(*)}$$

Since $\operatorname{codim}_X Y = 1$, (1) implies that

(2) $$SS(\mathfrak{M}) \cap P_Y^* X = \emptyset\ ,$$

namely, Y is non-characteristic with respect to $\mathfrak{M}$. Note that (1) does not imply (2) if $\operatorname{codim}_X Y \not\geq 1$.

This non-characteristic condition guarantees that the tangential system $\mathfrak{M}_Y = \mathfrak{D}_{Y \to X} \otimes \mathfrak{M}$ of $\mathfrak{M}$ induced on N is well defined. (S-K-K [8] Chap. II. Theorem 3.5.3.) On the other hand, $S_N^* X$ is the disjoint union of G_+, G_- and $\sqrt{-1}\,S^*M \underset{M}{\times} N$, where $G_\pm = \{(x, (\xi + \sqrt{-1}\,\eta)\infty) \in S_N^* X;\ x_1 = 0,\ \xi = \pm(c, 0, \cdots, 0)\,(c > 0)\}$. Let us denote by ρ the canonical projection from $S_N^* X - S_Y^* X$ to $S_N^* Y = \sqrt{-1}\,S^*N$. Then (1) guarantees

(3) $$\mathfrak{N}_\pm = \rho_*(\mathcal{E}_{Y \to X} \otimes (\mathfrak{M}|_{G_\pm}))$$

is a well-defined $\mathcal{E}_Y$-Module. Furthermore it follows from the definition that

(4) $$\mathcal{E}_Y \otimes \mathfrak{M}_Y = \mathfrak{N}_+ \oplus \mathfrak{N}_-\ .$$

Now the hyperfunction solutions of $\mathfrak{M}$ supported by $Z_\pm$ are related to the microfunction solutions of $\mathfrak{N}_\pm$ by the following Theorem 1. Here

(*) For a system $\mathfrak{M}$ of linear differential equations $SS(\mathfrak{M})$ denotes $\operatorname{Supp}(\mathcal{E}_X \otimes \mathfrak{M})$, namely, the characteristic variety of $\mathfrak{M}$.

$\mathbf{R}\,\mathcal{H}om(\ ,\ )$ etc. means the right derived functor associated with $\mathcal{H}om(\ ,\ )$ etc. (See e.g. Hartshorne [0] for the theory of derived categories and derived functors.)

THEOREM 1. *We have the following isomorphism:*

$$\text{(5)}\qquad \mathbf{R}\Gamma_{Z_\pm}\mathbf{R}\,\mathcal{H}om_{\mathcal{D}_X}(\mathcal{M},\mathcal{B}_M) \xrightarrow{\sim} \mathbf{R}\pi_{N*}\mathbf{R}\,\mathcal{H}om_{\mathcal{E}_Y}(\mathcal{N}_\pm,\mathcal{C}_N)\ [-1]\,.$$

In order to explain the implication of the isomorphism (5) we consider the following special case (cf. Example 3 below):

(a) The system $\mathcal{M}$ is locally solvable near 0, namely, $\mathcal{E}xt^j_{\mathcal{D}_X}(\mathcal{M},\mathcal{B}_M) = 0$ holds near 0 for $j \neq 0$.

(b) π_N gives rise to an isomorphism between $\operatorname{Supp}\mathcal{N}_+$ (= the characteristic variety of $\mathcal{N}_+$) and N.

Note that the condition (a) is not restrictive, because it follows from the ellipticity of $\mathcal{M}$ that $\mathcal{E}xt^j_{\mathcal{D}_X}(\mathcal{M},\mathcal{B}_M) \simeq \mathcal{E}xt^j_{\mathcal{D}_X}(\mathcal{M},\mathcal{A}_M)$(*) holds. Under these simplifying assumptions the isomorphism (5) entails

$$\text{(5}'\text{)}\qquad \mathcal{H}^j_{Z_+}(\mathcal{S}) \simeq \mathcal{E}xt^{j-1}_{\mathcal{E}_Y}(\mathcal{N}_+,\mathcal{C}_N)\,.$$

Here $\mathcal{S}$ denotes the (hyperfunction) solution sheaf of $\mathcal{M}$, namely, $\mathcal{E}xt^0_{\mathcal{D}_X}(\mathcal{M},\mathcal{B}_M)$. Furthermore

$$\mathcal{H}^j_{Z_+}(\mathcal{S})_0 \simeq \varinjlim_{U\ni 0} H^{j-1}(U\cap M_-,\mathcal{S}) \qquad (j\neq 1)$$

and $\mathcal{H}^1_{Z_+}(\mathcal{S})_0 \simeq \varinjlim_{U\ni 0} H^0(U\cap M_-,\mathcal{S})/H^0(U,\mathcal{S})$ hold thanks to the local solvability of the system $\mathcal{M}$. Here U runs over a fundamental system of open

(*) Here $\mathcal{A}_M$ denotes the sheaf of real analytic functions on M.

neighborhoods of 0. Hence the local structure of hyperfunction solutions of $\mathfrak{M}$ on $U \cap M_-$ is described in terms of the structure of microfunction solutions of $\mathcal{N}_+$. In particular, the obstruction against the local extensibility of solutions of $\mathfrak{M}$ across N from the side M_- is given by $\mathcal{E}xt^0_{\mathcal{E}_Y}(\mathcal{N}_+, \mathcal{C}_N)$.

See also Example 2 in [1] and Example 4 in [2] for concrete illustrations of our result.

As a simple and interesting example of applications of Theorem 1 we first mention the following example.

EXAMPLE 2. Let $f(t,x)$ be a real-valued real analytic function of $(t,x) \in \mathbf{R}^1 \times \mathbf{R}^{(n-1)}$ which satisfies the conditions

$$f(t,0) = 0 \tag{6}$$

and

$$\{(t,x) \in \mathbf{R}^n;\ \mathrm{grad}_x f(t,x) = 0\} = \{(t,x) \in \mathbf{R}^n;\ x = 0\}. \tag{7}$$

Define a system $\mathcal{N}_f$ by the following:

$$\mathcal{N}_f : \left(\left(1 + \sqrt{-1}\,\frac{\partial f}{\partial t}\right)\frac{\partial}{\partial x_j} - \sqrt{-1}\,\frac{\partial f}{\partial x_j}\,\frac{\partial}{\partial t}\right)u = 0, \qquad j = 1, \cdots, d.$$

We assume for the moment that $d = n-1$. Then, by choosing $\mathfrak{M}$ defined on $\mathbf{R}^2 \times \mathbf{R}^{(n-1)}$ by

$$\begin{cases} \dfrac{\partial}{\partial x_j} u = 0, & j = 1, \cdots, n-1 \\ \left(\dfrac{\partial}{\partial t} + \sqrt{-1}\,\dfrac{\partial}{\partial s}\right)u = 0 \end{cases}$$

and $N = \{(t,s,x) \in \mathbf{R}^{1+n};\ s = f(t,x)\}$, we can apply (5) to prove

$$R\,\mathcal{H}om(\mathcal{N}_f, \mathcal{C}_N)_{(0,\sqrt{-1}dt\infty)} \simeq R\Gamma_B(\mathbf{C}) \otimes \mathcal{C}_{\mathbf{R},(0,\sqrt{-1}dt\infty)}, \tag{8}$$

where $B = \{(t,x) \in \mathbf{R}^n;\ f(t,x) \geq 0\}$.

In other words, the structure of microfunction solutions of $\mathcal{N}_f$ at $(0, \sqrt{-1}\,dt\infty)$, namely $\mathcal{E}xt^j(\mathcal{N}_f, \mathcal{C}_N)_{(0,\sqrt{-1}\,dt\infty)}$, is described by the topological structure of B, namely $\mathcal{H}^j_B(C)_0$. Hence one might call this result to be a "Morse-theory type result."

The interest of the system $\mathcal{N}_f$ lies in the fact that it is a canonical form of a system $\mathcal{M}$ which satisfies

(9) $SS(\mathcal{M})$ is a non-singular submanifold of codimension $d(=n-1)$

(10) $W = SS(\mathcal{M}) \cap \sqrt{-1}\,S^*M$ is a non-singular submanifold of $\sqrt{-1}\,S^*M$ of codimension $2d\,(=2(n-1))$

and

(11) $\omega|_W$ provides W with the contact structure.

(See Sato-Kawai-Kashiwara [9] p. 233 for the proof.)

The isomorphism (8) can be obtained even when $d < n-1$ (by using the boundary value problems for higher codimensional boundary). However, we do not know whether $\mathcal{N}_f$ can be chosen as a canonical form of systems satisfying conditions (9), (10) and (11), except for another extreme case where $d = 1$. (See Sato-Kawai-Kashiwara [9]. See also Sato-Kawai-Kashiwara [10] for the case $d = 1$.) See also Treves [11] and Maire-Treves [7] for some related topics.

Next we discuss one of the most important examples of the isomorphism (5).

EXAMPLE 3. Let M be $C^n = R^{2n}$ and let N be $\{x = (z, \bar{z}) \in R^{2n};\ f(x) \equiv f(z, \bar{z}) = 0\}$. Here $f(z, \bar{z})$ is a real valued real analytic function such that $\mathrm{grad}_{(z,\bar{z})} f|_N \neq 0$. Define $M_\pm$ by $\{x \in M;\ \pm f(x) > 0\}$. Let X (resp. Y) be a complexification of M (resp. N). Let $\mathcal{M}$ be the Cauchy-Riemann equations on C^n, namely, $\mathcal{M} = \mathcal{D}_X \Big/ \left(\sum_{j=1}^{n} \mathcal{D}_X \bar{\partial}_j\right)$. Since the real locus of the characteristic variety of $\mathcal{M}_Y = \mathcal{D}_{Y \leftarrow X} \otimes \mathcal{M}$ can be identified with

$S^*_N M = N_+ \sqcup N_-$, the isomorphism (5) reads as follows in this case:

$$\mathcal{H}^k_{\underline{M}_\pm}(\mathcal{O}_M) \simeq \mathcal{E}xt^{k-1}_{\mathcal{E}_Y}(\mathcal{M}_Y, \mathcal{C}_N)|_{N_\pm} . \tag{12}$$

Here $N_\pm = \{(x, \sqrt{-1}\,\xi\infty) \in \sqrt{-1}\,S^*N;\ x \in N,\ \xi = \mathrm{grad}_z f\}$.

If we assume in addition that the signature of the generalized Levi form of $\mathcal{M}_Y$ is $(p, n-1-p)$ on N_+, then Theorem 2.3.6 of S-K-K [8] Chap. III asserts that

$$\mathcal{E}xt^{k-1}_{\mathcal{E}_Y}(\mathcal{M}_Y, \mathcal{C}_N)|_{N_+} = 0 \qquad \text{if } k \neq p \tag{13}$$

and

$$\mathcal{E}xt^{p-1}_{\mathcal{E}_Y}(\mathcal{M}_Y, \mathcal{C}_N)|_{N_+} \simeq \mathcal{C}' \tag{14}$$

hold. Here $\mathcal{C}'$ is isomorphic to the sheaf of microfunctions defined on $\sqrt{-1}\,S^*R^n$ by a suitable quantized contact transformation.

Note also that the signature of the generalized Levi form of $\mathcal{M}_Y$ considered on N_+ is equal to the signature of the Levi form associated with N.

This example suggests that we might transform the study of the structure of microfunction solutions of micro-differential equations defined on $\sqrt{-1}\,S^*R^n$ to that of holomorphic solutions by making use of the isomorphism (12). This is actually the case, as seen in Kashiwara-Kawai [3], [4], Kashiwara-Kawai-Oshima [5], [6]. As one of the typical examples of such a study, we shall review the work of Kashiwara-Kawai-Oshima [5], [6].

We begin our discussions by clarifying the general mechanism. We first prepare the following notations:(*)

(13) $\mathcal{J}_N$ is the defining Ideal of $\mathcal{M}_Y$, i.e., $\mathcal{M}_Y = \mathcal{D}_Y/\mathcal{J}_N$.

(*) In the sequel we consider the problem in the framework of homogeneous symplectic geometry instead of contact geometry for convenience sake.

$$(14)\quad \mathcal{M}_0 = \mathcal{E}_{C^{2n-1}}/\mathcal{J}_0\,, \quad \text{where } \mathcal{J}_0 = \sum_{j=1}^{p} \mathcal{E}_{C^{2n-1}}\left(\frac{\partial}{\partial z_j} + \sqrt{-1}\, z_j \frac{\partial}{\partial z_{2n-1}}\right) + \sum_{j=p+1}^{n-1} \mathcal{E}_{C^{2n-1}}\left(\frac{\partial}{\partial z_j} - \sqrt{-1}\, z_j \frac{\partial}{\partial z_{2n-1}}\right)$$

$$(15)\qquad V = SS(\mathcal{M}_Y) = Y \underset{X}{\times} T^*X \hookrightarrow T^*Y$$

$$(16)\qquad V_0 = \operatorname{Supp} \mathcal{M}_0$$

$$(17)\qquad W_0 = V_0 \cap \sqrt{-1}\, T^*R^{2n-1}$$

$$(18)\qquad W_0^+ = \{(x, \sqrt{-1} < \eta, dx >) \in V_0 \cap \sqrt{-1}\, T^*R^{2n-1};\ \eta_{2n-1} > 0\}\,.$$

In the sequel we identify W_0^+ with an open set in $\{(x_n, \cdots, x_{2n-1}, \sqrt{-1}(\eta_n, \cdots, \eta_{2n-1})) \in \sqrt{-1}\, T^*L;\ \eta_{2n-1} > 0\}$, where $L \cong R^n$.

We denote its complexification $(= C^n)$ by L^C. Then we have the following result.

THEOREM 4. *Let* $N = \{z \in C^n;\ f(x) \equiv f(z, \bar{z}) = 0\}$ *be a non-singular real hypersurface of* C^n. *Assume that the Levi form of* N *is non-degenerate and its signature is* $(p, n-1-p)$. *Let* $p^* = (0, k\, d_x f)$ *be a point in* $T^*_N C^n$. *Let* ψ *be a homogeneous canonical transformation defined in a neighborhood of* p^* *in* T^*C^n *into* $T^*L^C = T^*C^n$ *such that* $\psi|_{T^*_N C^n}$ *defines an isomorphism from* $T^*_N C^n$ *to* $T^*_{R^n} C^n$ *in a neighborhood of* p^*. *Denote* $\psi(p^*)$ *by* q^*. *Let* Ψ *be an isomorphism from* $\mathcal{E}_{C^n, p^*}$ *onto* $\psi^{-1}\mathcal{E}_{L^C, q^*}$. *Then there exists an isomorphism*

$$(19)\qquad \tilde{\Psi} : \mathcal{H}^{p+1}_{\{z \in C^n;\ f \geq 0\}}(\mathcal{O}_{C^n})_{p^*} \simeq \psi^{-1}\mathcal{C}_{L, q^*}$$

such that the action of $\mathcal{D}_{C^n}$ *on* $\mathcal{H}^{p+1}_{\{z \in C^n;\ f \geq 0\}}(\mathcal{O}_{C^n})$ *is compatible with* Ψ *and* $\tilde{\Psi}$.

Proof. A basic result of S-K-K [8] (Chap. III Theorem 2.3.2) asserts that there exist a homogeneous canonical transformation ϕ from a complex neighborhood of $(0, k\,d_z f) \in T^*_N Y \subset T^*Y$ to a complex neighborhood of $(0, \sqrt{-1}\,k\,dx_{2n-1}) \in T^*_{R^{2n-1}} C^{2n-1} \subset T^*C^{2n-1}$, a C-Algebra isomorphism $\Phi : \mathcal{E}_{Y,(0,kd_z f)} \to \phi^{-1}\mathcal{E}_{C^{2n-1},(0,\sqrt{-1}\,k\,dx_{2n-1})}$, an isomorphism $\tilde{\Phi} : \mathcal{C}_{N,(0,kd_z f)} \to \phi^{-1}\mathcal{C}_{R^{2n-1},(0,\sqrt{-1}\,k\,dx_{2n-1})}$, which is compatible with Φ, and an isomorphism: $\mathcal{M}_Y \xrightarrow{\sim} \phi^{-1}\mathcal{M}_0$ induced from Φ. Hence there exists an isomorphism $\tilde{\tilde{\Phi}} : \mathcal{E}xt^{p-1}_{\mathcal{E}_Y}(\mathcal{M}_Y, \mathcal{C}_N)_{(0,kd_z f)} \xrightarrow{\sim} \phi^{-1}\mathcal{E}xt^{p-1}_{\mathcal{E}_{C^{2n-1}}}(\mathcal{M}_0, \mathcal{C}_{R^{2n-1}})_{(0,\sqrt{-1}\,k\,dx_{2n-1})}$, while $\mathcal{E}xt^{p-1}_{\mathcal{E}_{C^{2n-1}}}(\mathcal{M}_0, \mathcal{C}_{R^{2n-1}})|_{W_0^+}$ is isomorphic to $\mathcal{C}_L$. (S-K-K [8] Chap. III Theorem 2.3.6.) Since $\mathcal{E}_{LC}$ can be identified with $\mathcal{E}nd_{\mathcal{E}_{C^{2n-1}}}(\mathcal{M}_0)$, the action of $\mathcal{E}nd_{\mathcal{E}_Y}(\mathcal{M}_Y)$ on $\mathcal{E}xt^{p-1}_{\mathcal{E}_Y}(\mathcal{M}_Y, \mathcal{C}_N)|_{N_+}$ can be identified with the action of $\phi^{-1}\mathcal{E}_{LC} \simeq \phi^{-1}\mathcal{E}nd_{\mathcal{E}_{C^{2n-1}}}(\mathcal{M}_0)$ on $\phi^{-1}\mathcal{C}_L \simeq \phi^{-1}\mathcal{E}xt^{p-1}_{\mathcal{E}_{C^{2n-1}}}(\mathcal{M}_0, \mathcal{C}_{R^{2n-1}})|_{W_0^+}$. On the other hand, $\mathcal{E}nd_{\mathcal{E}_Y}(\mathcal{M}_Y)_{p^*} \simeq \mathcal{E}_{C^n,p^*}$. Thus we have an isomorphism $\Phi' : \mathcal{E}_{C^n,p^*} \xrightarrow{\sim} \phi^{-1}\mathcal{E}_{LC}$ which is compatible with $\tilde{\tilde{\Phi}}$. On the other hand, Φ' coincides with Ψ up to an inner automorphism (S-K-K [8] p. 429). Hence we may assume without loss of generality that $\Phi' = \Psi$.

Since there exists an isomorphism Θ from $\mathcal{D}_{C^n} = \mathcal{E}nd_{\mathcal{D}_X}(\mathcal{M})$ onto $\mathcal{E}nd_{\mathcal{D}_Y}(\mathcal{M}_Y)$, we can canonically transfer the action of $P(z, D_z) \in \mathcal{D}_{C^n}$ on $R\,\mathcal{H}om(\mathcal{M}, \mathcal{B}_M)$ (and hence on $R\Gamma_{\overline{M}_+} R\,\mathcal{H}om(\mathcal{M}, \mathcal{B}_M)$) to the action of $\Theta(P)$ on $R\,\mathcal{H}om_{\mathcal{E}_Y}(\mathcal{M}_Y, \mathcal{C}_N)|_{N_+}$ through the isomorphism (12). Hence choosing $\tilde{\Psi}$ to be the composite of the isomorphism (12) with $\tilde{\tilde{\Phi}}$, we obtain the required result. Q.E.D.

Since the homogeneous symplectic structure of $V \cap \sqrt{-1}\, T^*N$, namely, $\omega_Y|_{V\cap\sqrt{-1}T^*N}$, is identical with $\omega_{C^n}|_{V\cap\sqrt{-1}T^*N} = \omega_{C^n}|_{T^*_N C^n} = k d_z f(z, \bar{z})$ $(k \in R^\times)$, the following Theorem 5 is important in applying Theorem 4 to concrete problems. This theorem can be verified by a direct calculation.

THEOREM 5. *Let* X *and* Y *be analytic manifolds of dimension* n. *Let* Z *be a non-singular hypersurface of* $X \times Y$ *defined by* $\{(x, y) \in X \times Y;\ f(x, y) = 0\}$. *Assume that*

$$\begin{vmatrix} 0 & d_x f \\ d_y f & d_x d_y f \end{vmatrix} \neq 0 \text{ on } Z. \tag{20}$$

Then $T^*_Z(X\times Y) = \{(x, y, \xi, \eta) \in T^*(X\times Y);\ f(x, y) = 0, (\xi, \eta) = k\, \mathrm{grad}_{(x,y)} f(x, y)\}$ *is a homogeneous symplectic manifold with the canonical 1-form* $\omega = k d_x f$. *Furthermore the Poisson bracket* $\{\phi, \psi\}$ *of functions* $\phi(x, y, k)$ *and* $\psi(x, y, k)$ *on* $T^*_Z(X\times Y)$ *is given by the following formula*:

$$\{\phi, \psi\} = \left\{ \begin{vmatrix} 0 & k\dfrac{\partial\psi}{\partial k} & d_y\psi \\ k\dfrac{\partial\phi}{\partial k} & 0 & d_y f \\ d_x\phi & d_x f & d_x d_y f \end{vmatrix} - \begin{vmatrix} 0 & k\dfrac{\partial\phi}{\partial k} & d_y\phi \\ k\dfrac{\partial\psi}{\partial k} & 0 & d_y f \\ d_x\psi & d_x f & d_x d_y f \end{vmatrix} \right\} \Bigg/ k \begin{vmatrix} 0 & d_y f \\ d_x f & d_x d_y f \end{vmatrix} \tag{21}$$

REMARK. The condition (20) is necessary and sufficient for $T^*_Z(X \times Y)$ to become a homogeneous symplectic manifold with canonical 1-form $kd_x f(x, y)$.

This formula is effectively used to calculate the cohomology groups associated with a class of multiple characteristic operators by the aid of Theorem 4 (Kashiwara-Kawai-Oshima [5]). In order to simplify the presentation, we restrict ourselves to the single equation case. The results are as follows:

THEOREM 6. *Let* $\mathcal{M} = \mathcal{E}_X/\mathcal{E}_X P$ *be a micro-differential equation defined in a neighborhood of* $p^* \in \Lambda = T^*_M X$, *where* M *is a real analytic manifold and* X *is its complexification. Assume that* P *satisfies the following conditions:*

P *has the form* $P_1 P_2 + Q$, *with* $\mathrm{ord}\, P_j = m_j\, (j = 1, 2)$ *and* $\mathrm{ord}\, Q \leq m_1 + m_2 - 1$ *so that the following conditions (22)-(27) are satisfied:*

$$\{\sigma(P_2), \sigma(P_1)\}|_{V_1 \cap V_2} \neq 0, \quad \textit{where } V_j = \sigma(P_j)^{-1}(0) \quad (j = 1, 2) \tag{22}$$

$$d\left(\frac{\sigma(Q)}{\{\sigma(P_2), \sigma(P_1)\}}\right) \wedge \omega|_{V_1 \cap V_2} \quad \textit{never vanishes.} \tag{23}$$

$$\{\sigma(P_1), \sigma(P_1)^c\} \neq 0, \qquad \{\sigma(P_2), \sigma(P_2)^c\} \neq 0 \ ^{(*)} \tag{24}$$

$$V_1 \cap V_1^c = V_2 \cap V_2^c \ (\equiv W) \ ^{(*)} \tag{25}$$

$$\kappa \equiv \left.\frac{\sigma(Q)}{\{\sigma(P_2), \sigma(P_1)\}}\right|_W \quad \textit{never attains integral values.} \tag{26}$$

(*) For a holomorphic function f(p) defined on a complexification Λ^C of a real analytic manifold Λ, we denote by $f^c(p)$, the complex conjugate of f(p), namely, a holomorphic function defined on Λ^C which coincides with $\overline{f(p)}$ on Λ. The complex conjugate V^c of an analytic variety $V \subset \Lambda^C$ is, by definition, the variety defined by the complex conjugate of holomorphic functions defining V.

(27) $d\kappa$, $d\kappa^c$ *and* $\omega|_W$ *are linearly independent.*

Then we have the following result on the structure of $\mathcal{E}xt^j_{\mathcal{E}_X}(\mathcal{M}, \mathcal{C}_M)$ *near* p^* *classified according to the sign of* $\{\sigma(P_j), \sigma(P_j)^c\}$ $(j = 1, 2)$.

Case A: $\{\sigma(P_1), \sigma(P_1)^c\}(p^*)\{\sigma(P_2), \sigma(P_2)^c\}(p^*) < 0$

(28) $$\mathcal{E}xt^j_{\mathcal{E}_X}(\mathcal{M}, \mathcal{C}_M) = 0 \quad \textit{for any } j .$$

Case B(+): $\{\sigma(P_1), \sigma(P_1)^c\}(p^*), \{\sigma(P_2), \sigma(P_2)^c\}(p^*) > 0$

(29) $$\begin{cases} \mathcal{E}xt^j_{\mathcal{E}_X}(\mathcal{M}, \mathcal{C}_M) = 0, & j \neq 0 \\ \mathcal{E}xt^0_{\mathcal{E}_X}(\mathcal{M}, \mathcal{C}_M) \simeq \mathcal{C}^2_W \; ^{(*)} \end{cases}$$

Case B(−): $\{\sigma(P_1), \sigma(P_1)^c\}(p^*), \{\sigma(P_2), \sigma(P_2)^c\}(p^*) < 0$

(30) $$\begin{cases} \mathcal{E}xt^j_{\mathcal{E}_X}(\mathcal{M}, \mathcal{C}_M) = 0, & j \neq 1 \\ \mathcal{E}xt^1_{\mathcal{E}_X}(\mathcal{M}, \mathcal{C}_M) \simeq \mathcal{C}^2_W . \end{cases}$$

Theorem 6 asserts that both regularity theorem and existence theorem hold micro-locally for the micro-differential equation $Pu = f$ in Case A, while only the existence (resp. regularity) theorem holds in Case B(+) (resp. B(−)). Furthermore it claims that the obstruction against the solvability of $Pu = f$ is described by a flabby sheaf on W.

(*) Here and in the sequel, $\mathcal{C}_W$ (resp. $\mathcal{E}_W$) denotes the sheaf of micro-functions (resp. micro-differential operators) defined on an open set of $\sqrt{-1}\,T^*\mathbf{R}^n$ which is isomorphic to W with the homogeneous canonical 1-form $\omega|_W$.

Next let us investigate the structure of $\mathcal{E}xt^j_{\mathcal{E}_X}(\mathfrak{N}, \mathcal{C}_M)$ when κ attains integral values. For this purpose we first note that micro-differential operator R on W such that $\sigma(R) = \kappa$ is uniquely determined up to inner automorphisms of $\mathcal{E}_W$ (*) by the condition (23) (S-K-K [8] Chap. II. Th. 2.1.2). Hence $\mathcal{L}_\ell = \mathcal{E}_W/\mathcal{E}_W(R-\ell)$ $(\ell \in \mathbb{C})$ is well defined by κ up to isomorphisms. Actually, $\mathcal{E}xt^j_{\mathcal{E}_X}(\mathfrak{N}, \mathcal{C}_M)$ is described in terms of these as follows:

THEOREM 7. *Assume the same conditions as in Theorem 5 except for the condition* (26). *Then we have the following isomorphisms near* p^* :

Case A_1: $\{\sigma(P_1), \sigma(P_1)^c\}(p^*) < 0$, $\{\sigma(P_2), \sigma(P_2)^c\}(p^*) > 0$

$$\mathcal{E}xt^j_{\mathcal{E}_X}(\mathfrak{N}, \mathcal{C}_M) \simeq \bigoplus_{\ell=0,-1,-2,\cdots} \mathcal{E}xt^j_{\mathcal{E}_W}(\mathcal{L}_\ell, \mathcal{C}_W). \tag{31}$$

Case A_2: $\{\sigma(P_1), \sigma(P_1)^c\}(p^*) > 0$, $\{\sigma(P_2), \sigma(P_2)^c\}(p^*) < 0$

$$\mathcal{E}xt^j_{\mathcal{E}_X}(\mathfrak{N}, \mathcal{C}_M) \simeq \bigoplus_{\ell=1,2,3,\cdots} \mathcal{E}xt^j_{\mathcal{E}_W}(\mathcal{L}_\ell, \mathcal{C}_W). \tag{32}$$

Case B(+): $\{\sigma(P_1), \sigma(P_1)^c\}(p^*) > 0$, $\{\sigma(P_2), \sigma(P_2)^c\}(p^*) > 0$

$$\begin{aligned} 0 \to \bigoplus_{\ell=0,-1,-2,\cdots} \mathcal{E}xt^0_{\mathcal{E}_W}(\mathcal{L}_\ell, \mathcal{C}_W) \to \mathcal{C}_W^2 \\ \to \mathcal{E}xt^0_{\mathcal{E}_X}(\mathfrak{N}, \mathcal{C}_M) \to \bigoplus_{\ell=0,-1,-2,\cdots} \mathcal{E}xt^1_{\mathcal{E}_W}(\mathcal{L}_\ell, \mathcal{C}_W) \to 0 \ (exact) \end{aligned} \tag{33}$$

Case B(−): $\{\sigma(P_1), \sigma(P_1)^c\}(p^*) < 0$, $\{\sigma(P_2), \sigma(P_2)^c\}(p^*) < 0$

(*)The same as the footnote on p. 50.

$$(34)\quad 0 \to \mathcal{E}xt^0_{\mathcal{E}_X}(\mathcal{M}, \mathcal{C}_M) \to \bigoplus_{\ell=0,-1,-2,\cdots} \mathcal{E}xt^0_{\mathcal{E}_W}(\mathcal{L}_\ell, \mathcal{C}_W)$$

$$\to \mathcal{C}^2_W \to \mathcal{E}xt^1_{\mathcal{E}_X}(\mathcal{M}, \mathcal{C}_M) \to \bigoplus_{\ell=0,-1,-2,\cdots} \mathcal{E}xt^1_{\mathcal{E}_W}(\mathcal{L}_\ell, \mathcal{C}_W) \to 0 \quad (exact).$$

REMARK. We want to emphasize the following two interesting points of the above results. Especially the second point seems to us to be very important. For the sake of definiteness, we concentrate our attention on Case A_1.

First, both the regularity theorem and the existence theorem hold for the equation in question even if κ attains strictly positive integers.

Second, in case κ attains 0 or strictly negative integers, say k, the structure of $\mathcal{E}xt^j_{\mathcal{E}_X}(\mathcal{M}, \mathcal{C}_M)$ is controlled by the micro-differential equation $(R-k)u = 0$ on W.

We shall give a sketch of the proof of these results in what follows:

The first step is to reduce the operator to a simpler form on $\Lambda^{\mathbf{C}}$. By a suitable quantized contact transformation, we can easily see that $\mathcal{M}$ is isomorphic to $\mathcal{E}_{\mathbf{C}^n}/\mathcal{E}_{\mathbf{C}^n}(z_1D_1 - z_2)$ considered near the point $(z,\zeta) \in T^*\mathbf{C}^n$, where $z_1 = \zeta_1 = 0$. Here we have used conditions (22) and (23). Again using another quantized contact transformation that preserves $z_1D_1 - z_2$, we can find a real non-singular hypersurface N so that

$$(35)\qquad T^*_N U = \Lambda \subset \Lambda^{\mathbf{C}} = T^*U$$

holds for an open neighborhood U of 0 in $\mathbf{C}^n$. Actually, such a fibering of $\Lambda^{\mathbf{C}}$ over $\mathbf{C}^n$ can be found as follows:

First we consider the problem infinitesimally. For this purpose we define a skew symmetric bilinear form $E(v_1, v_2)$ on $S \equiv T_{p^*}(T^*X)$ by $\langle d\omega, v_1 \wedge v_2 \rangle$. Define $S_{\mathbf{R}}$ by $T_{p^*}(\sqrt{-1}\,T^*M)$. For an $\mathbf{R}$-linear subspace T of S, we denote by $T^\perp$ the orthogonal complement of T with respect to Re E, i.e. $T^{\perp} = \{v \in S;\ \mathrm{Re}\, E(v, w) = 0$ holds for any $w \in T\}$. For a $\mathbf{C}$-linear subspace ρ of S such that $\rho \subset \rho^{\perp}$, T^{ρ} denotes $[(T \cap \rho^{\perp}) + \rho]/\rho$. Note that $\rho^{\perp}/\rho$ naturally becomes a symplectic vector space by E. We now define a linear form $f_j (j = 1, 2)$ on S by $d\sigma(P_j)(p^*)$. We also define a linear form f_3 on S by

$$d\sigma(Q)(p^*) - \frac{\{\sigma(P_2), \sigma(Q)\}(p^*)}{\{\sigma(P_2), \sigma(P_1)\}(p^*)} f_1 - \frac{\{\sigma(P_1), \sigma(Q)\}(p^*)}{\{\sigma(P_1), \sigma(P_2)\}(p^*)} f_2 .$$

Then we have

$$E(f_1, f_3) = E(f_2, f_3) = 0 . \tag{36}$$

Since the condition (25) implies that f_1^c and f_2^c belong to $\mathbf{C}f_1 + \mathbf{C}f_2$, we have also

$$E(f_1^c, f_3) = E(f_2^c, f_3) = 0 . \tag{37}$$

In what follows we denote $f_j^{-1}(0)$ by T_j.

In the sequel, we identify S with $S^* (= T^*_{p^*}(T^*X))$ by assigning $f \in S^*$ to $v \in S$ so that $E(w, v) = f(w)$ holds for any $w \in S$. By this identification $T_j^{\perp} = \mathbf{C}f_j$ $(j = 1, 2, 3)$ holds. We define ρ_0 by $\mathbf{R}\ \mathrm{Re}\, \omega (= (\mathrm{Re}\, \omega = 0)^{\perp})$ and ρ by $\mathbf{C}f_1 + \mathbf{C}f_3 + \mathbf{C}\rho_0$. Clearly ρ is isotropic. Then it follows from the conditions (24), (25) and (27) that

$$\rho \cap S_{\mathbf{R}} = \rho_0 . \tag{38}$$

In fact, if we assume

$$af_1 + bf_3 + c\omega = a'f_1^c + b'f_3^c + c'\omega \qquad (a, b, c, a', b', c' \in \mathbf{C}) ,$$

then we have $a = a' = 0$ by (24), (36), (37) by operating $E(f_1, *)$ and $E(f_1^c, *)$ to the above equality. Then (27) assures that $b = b' = 0$. Hence we get (38).

Next we show that there exists a Lagrangian $\mathbf{C}$-subspace λ of S (i.e., $\lambda^{\perp} = \lambda$) such that

$$\lambda \supset \rho \tag{39}$$

and

$$\lambda \cap S_{\mathbf{R}} = \rho \cap S_{\mathbf{R}} \tag{40}$$

hold.

Since $(S_{\mathbf{R}}^{\rho})^{\perp} = (S_{\mathbf{R}}^{\perp})^{\rho} = S_{\mathbf{R}}^{\rho}$ holds, $S_{\mathbf{R}}^{\rho} \subset \rho^{\perp}/\rho$ is Lagrangian. Hence we can find a Lagrangian $\mathbf{C}$-subspace $\vartheta \subset \rho^{\perp}/\rho$ such that

$$\vartheta \cap S_{\mathbf{R}}^{\rho} = 0 \tag{41}$$

holds. Let λ be a $\mathbf{C}$-subspace of $\rho^{\perp}$ such that $\vartheta = \lambda/\rho$ holds. Then it is easy to see that λ is Lagrangian in V.

On the other hand, (41) implies

$$\lambda \cap [(S_{\mathbf{R}} + \rho) \cap \rho^{\perp}] = \rho \tag{42}$$

holds. Since λ is contained in $\rho^{\perp}$, (42) entails that $\lambda \cap S_{\mathbf{R}} \subset \rho$ holds. Hence we find a Lagrangian subspace λ which satisfies both (39) and (40). Using this subspace λ as an "initial condition," we will construct the required fibering of $\Lambda^{\mathbf{C}} = T^*\mathbf{C}^n$. We denote the fiber coordinate of $T^*\mathbf{C}^n$ by ζ.

First, since $f_1 \in \lambda$, we can define ϕ_1 which is homogeneous of degree 0 with respect to ζ so that $V_1 = \{\phi_1 = 0\}$ and that $f_1 = d\phi_1(p^*)$. Secondly we determine ψ_1 on $\Lambda^{\mathbf{C}}$ by solving

$$\begin{cases} \{\psi_1, \phi_1\} = 1 \\ \psi_1|_{V_2} = 0 \,. \end{cases}$$

Then ψ_1 is necessarily homogeneous of degree 1 with respect to ζ, and $d\psi_1(p^*) = f_2/E(f_1, f_2)$.

Thirdly we define ϕ_2 on $\Lambda^{\mathbf{C}}$ by solving

$$(43)\quad \begin{cases} \{\psi_1, \phi_2\} = \{\phi_1, \phi_2\} = 0 & (43.a) \\ \phi_2|_{V_1 \cap V_2} = \kappa & (43.b) \end{cases}$$

Since κ is homogeneous of degree 0 with respect to ζ, ϕ_2 is also homogeneous of degree 0 with respect to ζ. Furthermore (43.a) implies that $E(f_2, d\phi_2(p^*)) = E(f_1, d\phi_2(p^*)) = 0$. On the other hand, (43.b) entails that $d\phi_2(p^*)$ is contained in $\mathbf{C}f_1 + \mathbf{C}f_2 + \mathbf{C}f_3$. Hence, together with (36), we conclude that $d\phi_2(p^*) \in \mathbf{C}f_3$. Therefore $d\phi_2(p^*) \in \lambda$.

Since λ is Lagrangian, we can further determine $\psi_j (j = 2, \cdots, n)$, which is homogeneous of degree 1 with respect to ζ, and $\phi_j (j = 3, \cdots, n)$, which is homogeneous of degree 0 with respect to ζ, by solving the following equations:

$$(44)\quad \begin{cases} \{\psi_j, \psi_k\} = \{\phi_j, \phi_k\} = 0 & (1 \le j, k \le n) \\ \{\psi_j, \phi_k\} = \delta_{jk} & (1 \le j, k \le n) \end{cases}$$

$$(45)\quad d\phi_j(p^*) \in \lambda \qquad (1 \le j \le n).$$

The canonical coordinate system $(\phi_1, \cdots, \phi_n, \psi_1, \cdots, \psi_n)$ thus constructed provides us with a homogeneous canonical transformation from $\Lambda^{\mathbf{C}}$ to $T^*\mathbf{C}^n$. Since $\lambda \cap S_{\mathbf{R}} = \mathbf{R}\,\mathrm{Re}\,\omega$,

$$(46)\quad T_{p^*}\Lambda \to T_0\mathbf{C}^n$$

has rank $(2n-1)$, which implies that $\Lambda \to \mathbf{C}^n$ has rank $2n-1$ in a neighborhood of p^*.

This implies that Λ is a conormal bundle of a real non-singular hypersurface N of U, where U is an open neighborhood of 0 of $\mathbf{C}^n$.(*)

(*)For the sake of the simplicity of the subsequent arguments, we assume that U is a polydisc.

Thus we have constructed the required fibering of $\Lambda^{\mathbb{C}}$ over U. Let $f(z, \bar{z})$ be a defining function of N. Since $\omega|_\Lambda = k d_z f(z, \bar{z})$ $(k > 0)$,

$$\begin{vmatrix} 0 & d_{\bar{z}} f(z, \bar{z}) \\ {}^t d_z f(z, \bar{z}) & d_z d_{\bar{z}} f \end{vmatrix} \neq 0 . \tag{47}$$

holds on N. In particular, this implies that the Levi form of N is non-degenerate. Hence Theorem 4 is applicable to our problem. Therefore it suffices for us to calculate $R\Gamma_{Z_+} R\mathcal{H}om_{\mathcal{D}_{\mathbb{C}^n}}(\mathcal{L}, \mathcal{O}_{\mathbb{C}^n})_0[-1]$ to calculate $R\mathcal{H}om_{\mathcal{E}_X}(\mathcal{N}, \mathcal{C}_M)_{p^*}$, where $\mathcal{L} = \mathcal{D}_{\mathbb{C}^n}/\mathcal{D}_{\mathbb{C}^n}(z_1 D_1 - z_2)$ and $Z_+ = \{z \in U; f(z, \bar{z}) \geq 0\}$ in the coordinate system chosen above. For this purpose we rewrite conditions (24) and (25) in terms of f.

First of all, (25) reads

$$\{z_1 = \bar{z}_1 = 0, f(z, \bar{z}) = 0\} = \left\{\frac{\partial f}{\partial z_1} = \frac{\partial f}{\partial \bar{z}_1} = 0, f(z, \bar{z}) = 0\right\} . \tag{48}$$

We may assume without loss of generality that f has the form

$$z_n + \bar{z}_n + \phi(z) z_1 + \overline{\phi(z)} \bar{z}_1 + \psi(z) \bar{z}_1 + \overline{\psi(z)} z_1 + \Phi(z'', z_n, \bar{z}'', \bar{z}_n) + O(|z|^3) , \tag{49}$$

where $z'' = (z_2, \cdots, z_{n-1})$, ϕ and ψ are linear and Φ is a quadratic form. Define $\phi_0(z'')$ and $\alpha \in \mathbb{C}$ (resp. $\psi_0(z'')$ and $\beta \in \mathbb{C}$) by $\phi(0, z'', z_n) = \phi_0(z'') + \alpha z_n$ (resp. $\psi(0, z'', z_n) = \psi_0(z'') + \beta z_n$). Then comparing the tangent spaces at the origin of the both hand sides of (48) we find that $z_n + \bar{z}_n = 0$ entails

$$\begin{cases} \phi_0(z'') + \alpha z_n + \bar{\beta} \bar{z}_n + \overline{\psi_0(z'')} = 0 \\ \overline{\phi_0(z'')} + \bar{\alpha} \bar{z}_n + \beta z_n + \psi_0(z'') = 0 \\ z_n + \bar{z}_n = 0 . \end{cases} \tag{50}$$

Setting $z_n = 0$ in (50), we obtain

$$\phi_0(z'') + \overline{\psi_0(z'')} = 0 , \tag{51}$$

and hence

$$\phi_0(z'') = \psi_0(z'') = 0 . \tag{52}$$

It follows then that $\alpha = \overline{\beta}$. Therefore f has the form

$$(53)\ z_n + \overline{z}_n + az_1\overline{z}_1 + bz_1z_1 + \overline{b}\overline{z}_1\overline{z}_1 + (z_n+\overline{z}_n)(\alpha z_1 + \overline{\alpha}\,\overline{z}_1) + \Phi(z'', z_n, \overline{z}'', \overline{z}_n) + O(|z|^3).$$

This means that $\tilde{f} = f/(1+\alpha z_1 + \overline{\alpha}\,\overline{z}_1)$ can be chosen to be a defining function of N which satisfies the additional conditions

$$(54)\quad \begin{cases} \left.\dfrac{\partial^2\tilde{f}}{\partial z_1 \partial z_j}\right|_{(z,\overline{z})=(0,0)} = 0, & j = 2, \cdots, n \\[2ex] \left.\dfrac{\partial^2\tilde{f}}{\partial z_1 \partial \overline{z}_j}\right|_{(z,\overline{z})=(0,0)} = 0, & j = 2, \cdots, n . \end{cases}$$

We denote this $\tilde{f}$ by f in the sequel. Then, by the aid of Theorem 5, we can easily find the following:

$$(55)\quad \begin{cases} \{z_1, z_1^c\}(0) = -\dfrac{L_{n-1}(0)}{kL_n(0)} \\[2ex] \{\zeta_1, \zeta_1^c\}(0) = -\dfrac{k\sigma(0)L_{n-1}(0)}{L_n(0)} , \end{cases}$$

where

$$L_n = \begin{vmatrix} 0 & d_{\overline{z}}f \\ \\ d_z f & d_z d_{\overline{z}} f \end{vmatrix} ,$$

$$L_{n-1} = \begin{vmatrix} 0 & \frac{\partial f}{\partial \bar{z}_2}, & \cdots, & \frac{\partial f}{\partial \bar{z}_n} \\ \frac{\partial f}{\partial z_2} & & & \\ \cdot & & \left(\frac{\partial^2 f}{\partial z_j \partial \bar{z}_k}\right)_{2 \le j,k \le n} & \\ \cdot & & & \\ \cdot & & & \\ \frac{\partial f}{\partial z_n} & & & \end{vmatrix}$$

and $\sigma = \dfrac{\partial^2 f}{\partial z_1^2} \dfrac{\partial^2 f}{\partial \bar{z}_1^2} - \left(\dfrac{\partial^2 f}{\partial z_1 \partial \bar{z}_1}\right)^2$.

Even though several cases are distinguished according to the signs of $\{\sigma(P_j), \sigma(P_j)^c\}$ $(j = 1, 2)$ in Theorems 6 and 7, the way of the discussion is the same. So we consider only one case here, say case (A_1).

First define $\phi(z', \bar{z}')$ by $f(z, \bar{z})|_{z_1 = \bar{z}_1 = 0}$, where $z' = (z_2, \cdots, z_n)$. Note that the condition (25) (hence the condition (48)) asserts that

$$C \equiv \left\{ z \in U;\ f(z, \bar{z}) = 0,\ \frac{\partial f}{\partial z_1} = \frac{\partial f}{\partial \bar{z}_1} = 0 \right\}$$

coincides with $\{z = (z_1, z') \in U;\ z_1 = 0, \phi(z', \bar{z}') = 0\}$. In the sequel we denote by F the projection from $\mathbf{C}^n_z$ to $\mathbf{C}^{n-1}_{z'}$. (See Figure 1.)

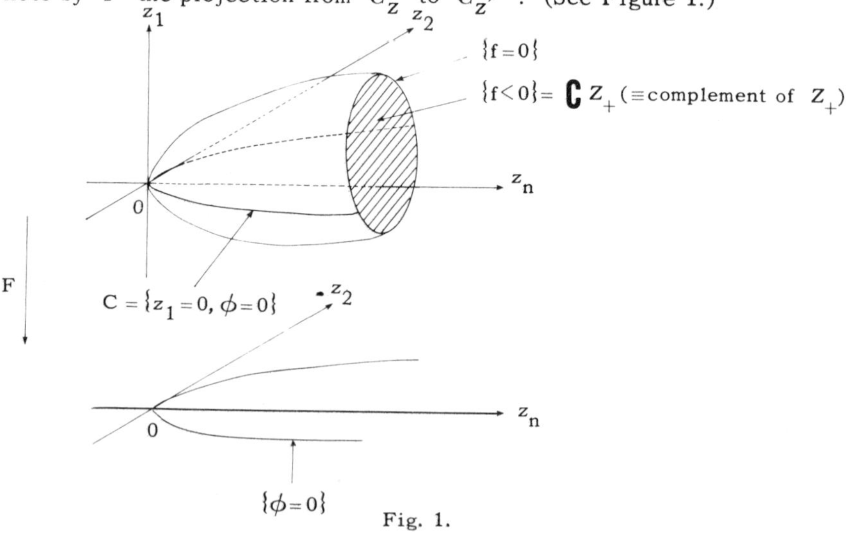

Fig. 1.

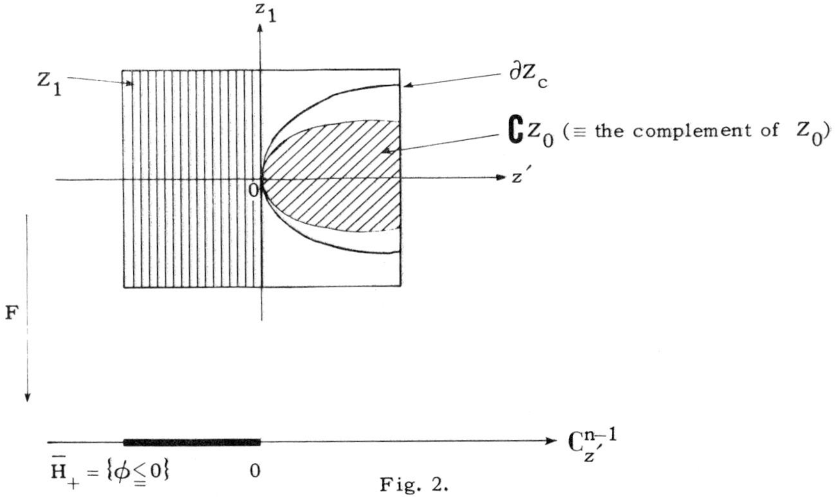

Fig. 2.

Note that in this case (i.e., Case A_1), $f(z, \bar{z})$ has the following form

(56) $$a z_1 \bar{z}_1 + b z_1^2 + \bar{b} \bar{z}_1^2 + \phi(z', \bar{z}') + O(|z_1|\,|z'|^2 + |z_1|^2 |z'|)\,,$$

where $a > 0$ and $a^2 > |b|^2$.

We next choose a family of closed sets $\{Z_c\}_{0 \leq c \leq 1}$ so that the following conditions (57)-(60) are satisfied. (See Figure 2 above.)

(57) $$Z_0 = Z_+ \text{ and } Z_1 = \{z \in F^{-1}(U') \cap U;\ \phi(z', \bar{z}') \geq 0\}\,,$$

where U' is an open ball in $C^{n-1}_{z'}$.

(58) $$\bigcap_{c' < c} Z_{c'} = Z_c$$

(59) $$F|_{Z_c} \text{ is proper.}$$

(60) $\partial Z_c\,(c < 1)$ is non-singular and non-characteristic with respect to $\mathfrak{L}$ outside C.

In the sequel we denote by $\bar{H}_+$ the set $\{z' \in U';\ \phi(z', \bar{z}') \geq 0\}$. We also define U'_ε (resp. $U_{\delta,\varepsilon}$) by $\{z' \in C^{n-1}, |z'| < \varepsilon\}$ (resp. $\{z = (z_1, z') \in C^n;$

$|z_1| < \delta, |z'| < \varepsilon\}$ for $0 < \varepsilon << \delta$. We denote $F|_{U_{\delta,\varepsilon}}$ by $F_{\delta,\varepsilon}$ for short. Then we have

$$(61)\qquad RF_{\delta,\varepsilon *}(R\Gamma_{Z_+}(R\,\mathcal{H}om(\mathcal{L},\mathcal{O}_{C^n})))_0$$
$$\simeq RF_{\delta,\varepsilon *}(R\Gamma_{Z_c}(R\,\mathcal{H}om(\mathcal{L},\mathcal{O}_{C^n})))_0$$
$$\simeq RF_{\delta,\varepsilon *}(\varprojlim_{c\uparrow 1} R\Gamma_{Z_c}(R\,\mathcal{H}om(\mathcal{L},\mathcal{O}_{C^n})))_0$$
$$\simeq RF_{\delta,\varepsilon *}(R\Gamma_{Z_1}(R\,\mathcal{H}om(\mathcal{L},\mathcal{O}_{C^n})))_0\,.$$

On the other hand, it follows from the choice of Z_1 that

$$R\Gamma_{Z_1}(R\,\mathcal{H}om(\mathcal{L},\mathcal{O}_{C^n}))_0 = \varinjlim_{\delta,\varepsilon\downarrow 0} R\Gamma(U_\varepsilon;\, RF_{\delta,\varepsilon *}R\Gamma_{Z_1}R\,\mathcal{H}om(\mathcal{L},\mathcal{O}_{C^n}))$$
$$= \varinjlim_{\delta,\varepsilon\downarrow 0} R\Gamma(U_\varepsilon;\, R\Gamma_{\overline{H}_+}RF_{\delta,\varepsilon *}R\,\mathcal{H}om(\mathcal{L},\mathcal{O}_{C^n}))$$

Next by the direct calculation concerning the ordinary differential equation with parameters, i.e., $(z_1D_1 - z_2)u(z) = f(z)$, we find

$$(62)\qquad \varinjlim_{\delta,\varepsilon\downarrow 0} R\Gamma(U_\varepsilon;\, R\Gamma_{\overline{H}_+}RF_{\delta,\varepsilon *}R\,\mathcal{H}om(\mathcal{L},\mathcal{O}_{C^n}))$$
$$\simeq \bigoplus_{\ell=0,1,2,\cdots} R\Gamma_{\overline{H}_+}(\mathcal{O}_{C^{n-1}}/(z_2-\ell)\mathcal{O}_{C^{n-1}})_0[-1]\,.$$

On the other hand $L_{n-1}(0)/L_n(0) > 0$ in Case A_1. This implies that the number of positive eigenvalues of the Levi form associated with $\{z'\in U';\ \phi(z',\overline{z}')=0\}$ is decreased by one compared with that of $\{z\in U;\ f(z,\overline{z})=0\}$. Therefore (62) combined with (12) and (19) entails

$$(63)\ \mathcal{E}xt^j_{\mathcal{E}_X}(\mathcal{M},\mathcal{C}_M) \simeq \bigoplus_{\ell=0,1,2,\cdots} H^j(\mathcal{C}_W \xrightarrow{-R-\ell} \mathcal{C}_W) = \bigoplus_{\ell=0,-1,-2,\cdots} \mathcal{E}xt^j_{\mathcal{E}_W}(\mathcal{L}_\ell,\mathcal{C}_W)\,.$$

This completes the proof of Theorem 5 in Case A and Theorem 6 in Case A_1. Other cases can be proved in a similar way.

REFERENCES

[0] Hartshorne, R.: Residues and Duality. Lecture Notes in Math. No. 20, Springer, Berlin-Heidelberg-New York, 1966.

[1] Kashiwara, M. and T. Kawai: On the boundary value problem for elliptic system of linear differential equations. I. Proc. Japan Acad., *48*, 712-715 (1972).

[2] ________: *Ibid.* II. Proc. Japan Acad., *49*, 164-168 (1973).

[3] ________: Theory of elliptic boundary value problems and its applications. Sûrikaiseki-Kenkyûsho Kôkyûroku, No. 238, RIMS, Kyoto Univ., Kyoto, 1975, pp. 1-59. (In Japanese.)

[4] ________: Finiteness theorem for holonomic systems of micro-differential equations. Proc. Japan Acad., *52*, 341-343 (1976).

[5] Kashiwara, M., T. Kawai and T. Oshima: Structure of cohomology groups whose coefficients are microfunction solution sheaves of systems of pseudo-differential equations with multiple characteristics. I. Proc. Japan Acad., *50*, 420-425 (1974).

[6] ________: *Ibid.* II. Proc. Japan Acad. *50*, 549-550 (1974).

[7] Maire, H. M. and F. Treves: An article on subelliptic systems (in preparation).

[8] Sato, M., T. Kawai and M. Kashiwara: (Referred to as S-K-K [8]) Microfunctions and pseudo-differential equations. Lecture Notes in Math. No. 287, Springer, Berlin-Heidelberg-New York, pp. 265-529 (1973).

[9] ________: The theory of pseudo-differential equations in hyperfunction theory. Sûgaku, *25*, 213-238 (1973). (In Japanese.)

[10] ________: On the structure of single linear pseudo-differential equations. Proc. Japan Acad., *48*, 643-646 (1972).

[11] Treves, F.: Study of a model in the theory of complexes of pseudo-differential operators. Ann. of Math. *104*, 269-324 (1976).

A SZEGO-TYPE THEOREM FOR SYMMETRIC SPACES

Victor Guillemin

§0. *Acknowledgements*

This article extends some results of Widom [7] from rank one to higher rank symmetric spaces. The proof of Theorem 3, which is the main result of this paper, makes use of a beautiful trick devised by Widom to prove the analogous rank one result in [7]. I am grateful to him for letting me make free use of results from [7] (which was as yet unpublished at the time this article was being written) and for providing me with some inspiring discussions.

I would also like to thank Sigurdur Helgason for valuable advice concerning the material in §4.

§1. *A Szego theorem for Lie groups*

Let G be a compact semi-simple Lie group and H a Cartan subgroup of G. Let $\mathfrak{g}$ and $\mathfrak{h}$ be the Lie algebras of G and H. For $\beta' \epsilon\, \mathfrak{h}^*$ let $|\beta|$ be the norm of β with respect to the Killing form. From the Killing form we get an orthogonal projection of $\mathfrak{g}$ on $\mathfrak{h}$, and by duality an injection $i: \mathfrak{h}^* \to \mathfrak{g}^*$. G acts on $\mathfrak{g}^*$ by its co-adjoint action. Given $\gamma \,\epsilon\, \mathfrak{h}^*$ we will denote by O_γ the G-orbit through $i(\gamma)$ in $\mathfrak{g}^*$. On O_γ there is a unique G-invariant measure, μ_γ, with the property that $\mu_\gamma(O_\gamma) = 1$.

Seminar on Micro-Local Analysis
0-691-08228-6/79/00 0063-16 \$00.80/1 (cloth)
0-691-08232-4/79/00 0063-16 \$00.80/1 (paperback)
For copying information, see copyright page

Let $r: \mathfrak{g}^* \to \mathfrak{h}^*$ be the transpose of the inclusion map of $\mathfrak{h}$ into $\mathfrak{g}$ and let $\nu_\gamma = r_* \mu_\gamma$. We can now state the main result of this paper:

THEOREM 1. *Let* $W_1, W_2, \cdots$ *be a sequence of vector spaces on which* G *acts irreducibly with* $N_i = \dim W_i$ *tending to infinity. Let* β_k *be the maximal weight of* W_k. *Denote by* ν_k *the following measure on* h^*:

$$\nu_k(f) = \frac{1}{N_k} \sum f\left(\frac{\beta_k^{(s)}}{|\beta_k|}\right), \qquad \forall f \in C(h^*),$$

where $\beta_k^{(s)}$, $s = 1, \cdots, N_k$ *are the weights of the representation of* G *on* W_k *(counted with multiplicity). Then the following two assertions are equivalent*:

i) $\beta_k/|\beta_k|$ *tends to a limit as* k *tends to* ∞.

ii) ν_k *tends weakly to a limit as* k *tends to* ∞.

Moreover, if the limit of (i) *is* γ, *the limit of* (ii) *is the measure*, ν_γ, *defined above.*

Kostant proves the following theorem in [5]: ([5], Theorem 8.2).

THEOREM 2. *The support of the measure*, ν_γ, *is the convex hull in* $\mathfrak{h}^*$ *of the orbit of* γ *under the action of the Weyl group.*

It is well known that if an irreducible representation of G has maximal weight, β, then every weight of the representation is contained in the convex hull of the orbit of β under the action of the Weyl group. Theorems 1 and 2 tell us that, asymptotically, the converse of this result is true. We have depicted some possibilities for the support of ν_γ, in the case of SU(3), in the figure below:

REMARK. Let $\{\beta_k\}$ be the sequence $\beta_k = k\beta$, β being a fixed weight. In [1], Boutet de Monvel and Guillemin prove a somewhat stronger result

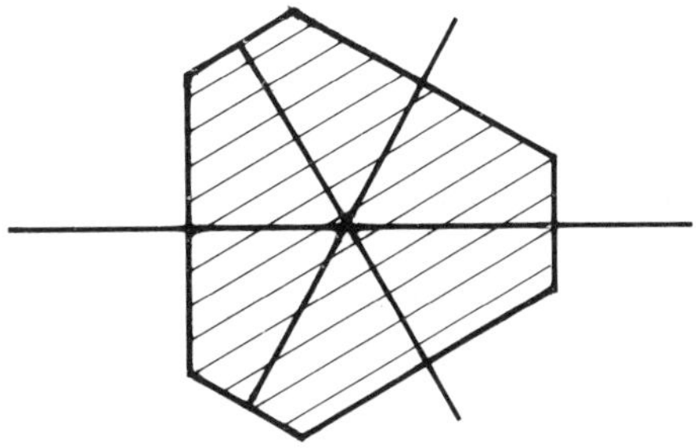

γ in the interior of the Weyl chamber

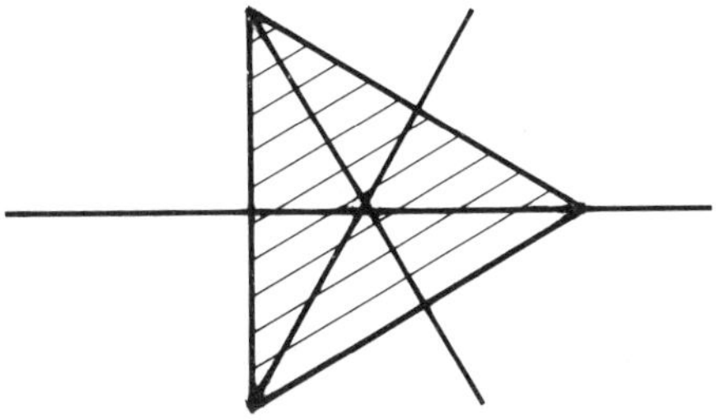

γ on the boundary of the Weyl chamber

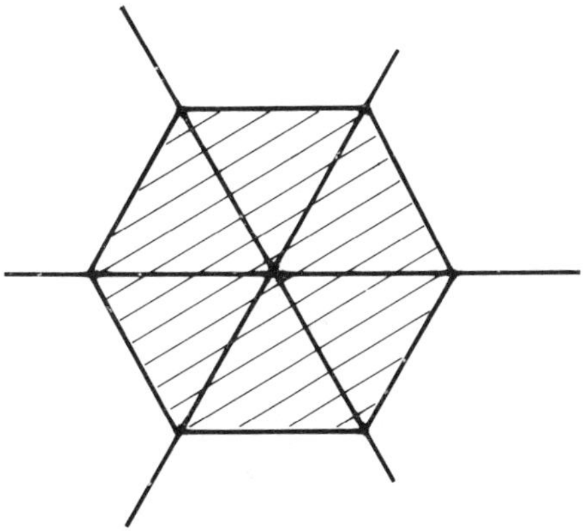

γ = the maximal weight of the adjoint representation.

than Theorem 1 for the associated sequence of ν_k's: Namely, for all $f \in C^\infty(\mathfrak{h}^*)$

$$\nu_k(f) = \sum_{r=0}^{\infty} \nu_\gamma^{(r)}(f) k^{-r}, \quad \text{with} \quad \nu_\gamma^{(0)} = \nu_\gamma ,$$

the $\nu_\gamma^{(r)}$'s being distributions on h^* supported on the convex hull of the orbit of γ under the Weyl group.

§2. *A Szego theorem for symmetric spaces*

Theorem 1 is a consequence of a more general result which we will formulate in this section. This result also generalizes Theorem 1 of Widom [7]. Let G be a compact semi-simple Lie group and let $X = G/K$ be a symmetric space. Let $\mathfrak{g}$ and $\mathfrak{k}$ be the Lie algebras of G and K and let $\mathfrak{p}$ be the orthogonal complement of $\mathfrak{k}$ in $\mathfrak{g}$ with respect to the Killing form. Let $\mathfrak{a}$ be a maximal abelian subalgebra of $\mathfrak{p}$ and $\mathfrak{h}$ a Cartan subalgebra of $\mathfrak{g}$ containing $\mathfrak{a}$. We can write $\mathfrak{h} = \tau \oplus \mathfrak{a}$ where $\tau = \mathfrak{h} \cap \mathfrak{k}$. It is well known that if V is a subspace of $L^2(X)$ on which G acts irreducibly and $\beta \in \mathfrak{h}^*$ is the maximal weight of the representation of G on V then $\beta(t) = 0$ for $t \in \tau$; so β is actually an element of $\mathfrak{a}^*$.

From the Killing form we get a projection of $\mathfrak{p}$ on $\mathfrak{a}$ and hence an injection, $\mathfrak{a}^* \to \mathfrak{p}^*$. Since $\mathfrak{p}^*$ is the cotangent space of X at the identity coset, o, an element, γ, of $\mathfrak{a}^*$ can be thought of as an element of T_o^*. The action of G on X lifts to an action of G on T^*X. We will denote by O_γ the G-orbit of $\gamma \in \mathfrak{a}^*$ in T^*X. As a homogeneous G-space O_γ possesses a unique G-invariant measure, μ_γ, with the property $\mu_\gamma(O_\gamma) = 1$.

THEOREM 3. *Let* $V_1, V_2, \cdots$ *be a sequence of subspaces of* $L^2(X)$ *on which* G *acts irreducibly with* $N_i = \dim V_i$ *tending to infinity. Let* β_i *be the maximal weight of the representation of* G *on* V_i, *and let* π_i *be the orthogonal projection of* $L^2(X)$ *on* V_i. *If* $\beta_i/|\beta_i|$ *tends to a limit,* γ, *in* $\mathfrak{a}^*$ *as* i *tends to infinity, then for every zeroth order pseudo-differential operator,* $B : L^2(X) \to L^2(X)$,

$$\lim_{k\to\infty} \frac{\operatorname{trace} \pi_k B \pi_k}{N_k} = \int_{O_\gamma} \sigma(B)\, d\mu_\gamma \tag{2.1}$$

where $\sigma(B)$ *is the symbol of* B *and* μ_γ *is the measure defined in the preceding paragraph.*

REMARKS.

1. In the course of the proof we will show that the left hand side of (2.1) has a limit for all B if and only if $\{\beta_i/|\beta_i|\}$ is convergent.

2. If X is of rank one, $\mathfrak{a}^*$ is one-dimensional. Since $\beta_i \in \mathfrak{a}^*_+$, the condition above on the sequence $\{\beta_i/|\beta_i|\}$ is trivially satisfied since $\beta_i/|\beta_i| = \gamma =$ the positive unit in $\mathfrak{a}^*$.

The proof of Theorem 3 requires some facts about generalized limits which we will describe in the next section and some results about invariant differential operators on X which we will discuss in §4.

§3. *Generalized limits*

Let ℓ^∞ be the Banach space of all bounded sequences $\mathfrak{c} = \{c_i\}$, $i \in Z^+$, with the norm $\|\mathfrak{c}\| = \sup|c_i|$. Let ℓ_0 be the closed subspace of ℓ^∞ consisting of all convergent sequences. The functional

$$L_0(\{c_i\}) = \lim c_i$$

is a continuous linear functional on ℓ_0. A *generalized limit* is a continuous linear functional, L, on ℓ^∞ whose restriction to ℓ_0 is L_0. The following lemma, which is an easy consequence of the Hahn-Banach theorem says that "sufficient many" generalized limits exist.

LEMMA 3.1. *Let* $\mathfrak{c}$ *be an element of* ℓ^∞. *If* $\mathfrak{c} \notin \ell_0$, *there exist generalized limits,* L_1 *and* L_2 *such that* $L_1(\mathfrak{c}) \neq L_2(\mathfrak{c})$.

Therefore, in order to show that a bounded sequence, $\mathfrak{c}$, is convergent it is enough to show that for all generalized limits, L, $L(\mathfrak{c})$ is a fixed number not depending on L.

§4. G-*invariant differential operators on* $X = G/K$

As in section 2 we will denote by $\mathfrak{g}$ and $\mathfrak{k}$ the Lie algebras of G and K, by $\mathfrak{p}$ the orthogonal complement of $\mathfrak{k}$ in $\mathfrak{g}$ and by $\mathfrak{a}$ a maximal abelian subalgebra of $\mathfrak{p}$. The orthogonal projection of $\mathfrak{p}$ on $\mathfrak{a}$ dualizes to an injection of $\mathfrak{a}^*$ into $\mathfrak{p}^*$. Also there is a canonical identification of $\mathfrak{p}^*$ with T_o^*, the cotangent space of X at the identity coset. Let us denote by $C_G^\infty T^*X$ the space of smooth G invariant functions on T^*X and by $C_K^\infty \mathfrak{p}^*$ the space of smooth functions on $\mathfrak{p}^*$ which are invariant under the co-adjoint action of K on $\mathfrak{p}^*$.

LEMMA 4.1. $$C_G^\infty T^*X \;\cong\; C_K^\infty \mathfrak{p}^* .$$

Proof. Since K is the isotropy subgroup of G at o, it acts linearly on T_o^*, and it is clear that under the identification of T_o^* with $\mathfrak{p}^*$ described above this action corresponds to the co-adjoint action of K on $\mathfrak{p}^*$. Thus if f is in $C_G^\infty T^*X$, its restriction to T_o^* is in $C_K^\infty \mathfrak{p}^*$. This gives us a map of $C_G^\infty T^*X$ into $C_K^\infty \mathfrak{p}^*$. It is easy to see it is bijective. Q.E.D.

Let M be the centralizer of $\mathfrak{a}$ in K and M' the normalizer of $\mathfrak{a}$ in K. Then M is of finite index in M' (see [2], page 244) and $W = M'/M$ acts on $\mathfrak{a}$ as a finite group of isometries. W is, by definition, the *Weyl group* of (G, K).

LEMMA 4.2. $$C_K^\infty \mathfrak{p}^* \;\cong\; C_W^\infty \mathfrak{a}^* .$$

Proof. Let $S_K\mathfrak{p}$ and $S_W\mathfrak{a}$ be, respectively, the rings of K-invariant polynomial functions on $\mathfrak{p}^*$ and W-invariant polynomial functions on $\mathfrak{a}^*$. It is clear that if f is a K-invariant function on $\mathfrak{p}^*$ its restriction to $\mathfrak{a}$ is W-invariant; so there are restriction maps, $S_K\mathfrak{p} \to S_W\mathfrak{a}$ and $C_K^\infty \mathfrak{p}^* \to C_W^\infty \mathfrak{a}^*$. In [2], it is shown that the first map is an isomorphism (see page 430); and it is also shown that every K-orbit in $\mathfrak{p}^*$ intersects $\mathfrak{a}^*$; so the second map is injective. By a theorem of G. Schwartz ([6]), every W-invariant C^∞ function, f, on $\mathfrak{a}^*$ can be written in the form, $f = F(g_1, \cdots, g_r)$ where F is smooth and $g_1, \cdots, g_r$ are in $S_W\mathfrak{a}$. Since the map from $S_K\mathfrak{p}$ to $S_W\mathfrak{a}$ is surjective, each g_i extends to a K-invariant function, g_i', on $\mathfrak{p}^*$; so f extends to the K-invariant function $f' = F(g_1', \cdots, g_r')$. Q.E.D.

Combining Lemmas 4.1 and 4.2 we get an isomorphism, given by restriction:

$$r: C_G^\infty T^*X \cong C_W^\infty \mathfrak{a}^* . \tag{4.1}$$

Let "$\| \quad \|$" be the G-invariant metric on T^*X whose restriction to $T_o = \mathfrak{p}$ is the Killing metric. Let S^*X be the unit co-sphere bundle of X and $\mathfrak{a}_1^*$ the unit cosphere in $\mathfrak{a}^*$.

LEMMA 4.3. *The restriction map is an isomorphism,* $r: C_G^\infty S^*X \to C_W^\infty \mathfrak{a}_1^*$.

Proof. The only non-obvious point is the surjectivity of r. Let ρ be a smooth function on the real line which is zero for $x < \frac{1}{2}$ and one for $x > \frac{3}{4}$. Given $f \in C_W^\infty \mathfrak{a}_1^*$, extend f by homogeneity to $\mathfrak{a}^* - 0$ and set $f_1(\xi) = \rho(|\xi|)f$. By (4.1) $\exists g_1 \in C_G^\infty T^*X$ such that $rg_1 = f_1$. Then $rg = f$ where $g = g_1 | S^*X$. Q.E.D.

LEMMA 4.3. *The map* r *preserves* sup *norms.*

Proof. Clearly $\sup|rg| \leq \sup|g|$ for each $g \in C_G^\infty S^*X$. Since every G-orbit in S^*X intersects $\mathfrak{a}_1^*$, this inequality is an equality. Q.E.D.

COROLLARY 4.4. r *extends to an isomorphism of Banach spaces*

$$r: C_G^0 S^*X \to C_W^0 \mathfrak{a}_1^* \tag{4.2}$$

and by duality to an isomorphism

(4.3) $$r^* : \mathfrak{M}_W \mathfrak{a}^* \to \mathfrak{M}_G S^*X$$

where $\mathfrak{M}_W \mathfrak{a}_1^*$ *is the space of* W-*invariant measures on* $\mathfrak{a}_1^*$ *and* $\mathfrak{M}_G S^*X$ *is the space of* G-*invariant measures on* S^*X.

Let us denote by $D_G X$ the ring of G-invariant differential operators on X. According to [2], (page 396), $D_G X$ is a commutative ring. Moreover, as an abstract ring, $D_G X$ can be identified with another, much simpler ring. Let $D_W \mathfrak{a}$ be the ring of constant coefficient W-invariant differential operators on the vector space, $\mathfrak{a}$. Then (*loc. cit.*, page 432) there is a canonical isomorphism

(4.4) $$\Psi : D_G X \cong D_W \mathfrak{a} .$$

We will not attempt to describe here how this isomorphism is constructed. We do need, however, one important property of it. (*Loc. cit.*, page 430.)

LEMMA 4.5. *The diagram*

(4.5) $$\begin{array}{ccc} D_G X & \xrightarrow{\ \Psi\ } & D_W \mathfrak{a} \\ \downarrow & & \downarrow \\ C_G^\infty T^*X & \xrightarrow{\ r\ } & C_W^\infty \mathfrak{a}^* \end{array}$$

commutes, the vertical arrows being the symbol maps.

We will also need later on the isomorphism

(4.6) $$D_W \mathfrak{a} \cong S_W \mathfrak{a}$$

given by the Fourier transform. (Recall that $S_W \mathfrak{a}$ is the ring of W-invariant polynomial functions on $\mathfrak{a}^*$.)

Let V be a subspace of $L^2(X)$ on which G acts irreducibly, and let P be an element of $D_G X$. P maps V into itself, and its restriction to V commutes with the action of G. Therefore, since the action of G on V is irreducible,

$$P = \chi_V(P) \text{ Identity}$$

on V. χ_V is a homomorphism of $D_G X$ into the complex numbers and is called the *infinitesmal character* of the representation of G on V. By (4.4) and (4.6) we get an associated homomorphism

$$\chi_V^{\#} : S_W \mathfrak{a} \to \mathbb{C} . \tag{4.7}$$

The following is an infinitesmal form of the Weyl character formula:

LEMMA 4.6. *Let* $\beta \in \mathfrak{a}^*$ *be the maximal weight of the representation of* G *on* V *and let* $\rho \in \mathfrak{a}^*$ *be the restriction to* $\mathfrak{a}^*$ *of* $\frac{1}{2}\sum \alpha_i$ *where the* α_i*'s are the positive roots of* $\mathfrak{g}$. *Then for every Weyl group invariant polynomial function on* $\mathfrak{a}^*$,

$$\chi_V^{\#}(p) = p(\beta + \rho) . \tag{4.8}$$

For the proof of this see [2], (X, §6) and [3].

§5. *The proof of Theorem 3*

The main ingredient in the proof is an argument due to Widom. (See [7], §1.) Let $V_1, V_2, \cdots$ be a sequence of subspaces of $L^2(X)$ satisfying the hypotheses of Theorem 3 and let π_i be the orthogonal projection of $L^2(X)$ on V_i. Let L be a generalized limit. If $N_i = \dim V_i$, then, for any bounded pseudodifferential operator, B, $(\text{trace } \pi_i B \pi_i) N_i^{-1}$ is uniformly bounded; so the generalized limit

$$L\left(\frac{\text{trace } \pi_i B \pi_i}{N_i}\right) \tag{5.1}$$

is defined. If B is a pseudodifferential operator of order < 0, it maps L^2 compactly into L^2; so $(\text{trace } \pi_i B \pi_i) N_i^{-1}$ tends to zero as i tends to infinity. This shows that for B of order zero, (5.1) is a function of the symbol of B alone. Moreover, if $\| \ \|$ is the supremum norm on L^2,

$$\sup_{T^*X-0} |\sigma(B)| = \inf\{\|B+K\|, K \text{ compact}\}$$

by a theorem of Kohn-Nirenberg ([4]); so (5.1) is *continuous* as a linear function of the symbol of B. By the Riesz representation theorem, there exists a measure, μ, on S^*X such that

$$L\left(\frac{\text{trace } \pi_i B \pi_i}{N_i}\right) = \int_{S^*X} \sigma(B)(x,\xi)\, d\mu \,. \tag{5.2}$$

This measure is a probability measure since the left hand of (5.2) is 1 when $B = I$. Given $g \in G$ let γ_g be the diffeomorphism induced on X by g. If B is a zeroth order pseudodifferential operator, the operator, B^g, defined by

$$B^g f = \gamma_g^* B (\gamma_g^{-1})^* f$$

is also a zeroth order pseudodifferential operator. Moreover,

$$\sigma(B)^g = \sigma(B^g)\,, \tag{5.3}$$

the left hand side of (5.3) being the symbol of B translated by g. Since

$$\text{trace } \pi_i B^g \pi_i = \text{trace } \gamma_g^* \pi_i B \pi_i (\gamma_g^{-1})^* = \text{trace } \pi_i B \pi_i$$

we get from (5.2) and (5.3)

$$\int_{S^*X} \sigma(B)^g \, d\mu = \int_{S^*X} \sigma(B)\, d\mu$$

for all zeroth order operators, B; i.e. for all smooth functions $\sigma(B)$ on S^*X. This shows that the measure, μ, is G-invariant. To determine μ, we will evaluate the left hand side of (5.2) on certain G-invariant zeroth order pseudodifferential operators.

LEMMA 5.1. *Let* $C: L^2(X) \to L^2(X)$ *be the operator which on each irreducible* G*-invariant subspace,* V, *of* $L^2(X)$ *takes the value,* $|\beta+\rho|$ Identity, β *being the maximal weight of the representation of* G *on* V *and* ρ *half the sum of the restricted roots. Then* C *is a positive, self-adjoint,* G*-invariant, first order elliptic pseudodifferential operator. Moreover, for all* $(x,\xi) \in T^*X - 0$,

$$\sigma(C)(x,\xi) = |\xi| . \tag{5.4}$$

Proof. Let Δ be the standard Laplace-Beltrami operator on the vector space, $\mathfrak{a}$. Clearly $\Delta \in D_W\mathfrak{a}$; so by Proposition 4.6, there is a G-invariant differential operator, Δ', on X with the property that on each irreducible subspace, V, of $L^2(X)$, $\Delta' = |\beta+\rho|^2$ Identity, β being the maximal weight. Moreover, by (4.5), the restriction of the symbol of Δ' to $\mathfrak{a}^*$ is equal to the symbol of Δ; i.e. on $\mathfrak{a}^*$, $\sigma(\Delta')(x,\xi) = |\xi|^2$. Since $\sigma(\Delta')$ is G-invariant, $\sigma(\Delta')(x,\xi) = |\xi|^2$ on all of T^*X. Set $C = (\Delta')^{1/2}$. Q.E.D.

Let q be an arbitrary homogeneous polynomial of degree k on $\mathfrak{a}^*$ which is Weyl group invariant, and let Q be the G-invariant differential operator on X associated with q via the identifications (4.4) and (4.6). Let $Q' = QC^{-k}$, a zeroth order G-invariant pseudodifferential operator. By Lemma (4.6)

$$\frac{\text{trace } \pi_i Q' \pi_i}{N_i} = \frac{q(\beta_i+\rho)}{|\beta_i+\rho|^k} = q\left(\frac{\beta_i+\rho}{|\beta_i+\rho|}\right) \tag{5.5}$$

where β_i is the maximal weight of the representation of G on V_i. This proves:

LEMMA 5.2. *The limit of the sequence* $\{(\text{trace } \pi_i Q'\pi_i)N_i^{-1}\}$ *exists for all* Q′ *of the form above if and only if the sequence* $\{\beta_i/|\beta_i|\}$ *is convergent. Moreover, if* $\{\beta_i/|\beta_i|\}$ *converges to the limit,* γ, *then* $\{(\text{trace } \pi_i Q'\pi_i)N_i^{-1}\}$ *converges to the limit* $\sigma(Q')(\gamma)$.

Now let us go back to the G-invariant measure, μ, occurring in the formula (5.2). By (4.5) this measure can be canonically identified with a Weyl group invariant measure, ν, on the unit sphere $\mathfrak{a}_1^* \subset \mathfrak{a}^*$. The space spanned by the functions, $q|\mathfrak{a}_1^*$, where q is a Weyl group invariant homogeneous polynomial on $\mathfrak{a}^*$, is dense in $C_W^0(\mathfrak{a}_1^*)$; so, by Lemma 5.2, the measure, ν, is just the delta measure at γ. Let O_γ be the orbit through γ in $T^*X - 0$. If a G-invariant function vanishes on O_γ its restriction to $\mathfrak{a}^*$ vanishes at γ; so the measure, μ, is concentrated on O_γ. Being a probability measure, it has to be the same as the measure, μ_γ, occurring in formula (2.1).

To recapitulate, we have proved that for *all* generalized limits, L:

$$L\left(\frac{\operatorname{trace} \pi_i B \pi_i}{N_i}\right) = \int_{O_\gamma} \sigma(B)\, d\mu_\gamma \,. \tag{5.6}$$

Therefore, by Lemma 3.1, the sequence $\{(\operatorname{trace} \pi_i B \pi_i) N_i^{-1}\}$ converges, and its limit is the right hand side of (5.6).

§6. *The proof of Theorem 1*

Let G be a compact semi-simple Lie group. Then $G \times G$ acts on G by the action, $(a,b)g = agb^{-1}$. The isotropy group at the identity element is G itself imbedded as the diagonal in $G \times G$. Therefore, we can think of G as a $(G \times G, G)$ homogeneous space, and it is not hard to show that this homogeneous space is in fact a symmetric space. (See [2], page 188.) In $L^2(G)$ we can consider either subspaces which are irreducible with respect to the right G-action or subspaces which are irreducible with respect to the $G \times G$ action. We will need

LEMMA 6.1. *If* V *is a* $G \times G$ *irreducible subspace,*

$$V = \bigoplus^{k} W, \quad k = \dim W \tag{6.1}$$

where W *is* G-*irreducible. Moreover, every irreducible representation of* G *occurs as such a* W.

Proof. See [8].

We will define on G a commutative algebra of zeroth order pseudo-differential operators as follows. First define $C: C^\infty(G) \to C^\infty(G)$ to be the operator which on each $G \times G$ –irreducible subspace, V, of $L^2(G)$ is equal to $|\beta + \rho|$ Identity, β being the maximal weight of the representation of G on W in (6.1). By Lemma 5.2, C is a positive, self-adjoint bi-invariant pseudodifferential operator of order one with symbol

$$\sigma(C)(x, \xi) = |\xi| \tag{6.2}$$

at all points $(x, \xi) \in T^*X$.

Let $\mathfrak{h}$ be a Cartan subalgebra of $\mathfrak{g}$. Given $f \in C^\infty(G)$ and $H \in \mathfrak{h}$, let $L_H f$ be the Lie derivative of f with respect to the left invariant vector field represented by H. Let D_H be the zeroth order pseudodifferential operator

$$(1/\sqrt{-1})\, L_H C^{-1} .$$

The D_H's are self-adjoint and commute among themselves; so they generate a commutative $*$-algebra, $\mathcal{H}$, of zeroth order left invariant pseudo-differential operators. If $H_1, \cdots, H_r$ are a basis of $\mathfrak{h}$, then $D_{H_1}, \cdots, D_{H_r}$ are a set of generators of $\mathcal{H}$. Moreover the map

$$S\mathfrak{h} \to \mathcal{H}, \quad p(\xi) \to p(D_H) \tag{6.3}$$

from the ring, $S\mathfrak{h}$, of polynomial functions on $\mathfrak{h}^*$ to $\mathcal{H}$ is a ring isomorphism. Let us compute the symbol of $p(D_H)$ for a given $p \in S\mathfrak{h}$. Since $p(D_H)$ is left-invariant, it is enough to compute the symbol at $(T_e^*)_1$, the unit sphere in T_e^*. Identify $(T_e^*)_1$ with the unit sphere, $\mathfrak{g}_1^*$, in $\mathfrak{g}^*$, and let $r: \mathfrak{g}^* \to \mathfrak{h}^*$ be the transpose of the inclusion map of $\mathfrak{h}$ into $\mathfrak{g}$.

LEMMA 6.2. *The symbol of* $p(D_H)$ *is the restriction to* $\mathfrak{g}_1^*$ *of the pull-back under* r *of the polynomial function,* p, *on* $\mathfrak{h}^*$.

Proof. It is enough to check this for the generators, $D_{H_i} = (1/\sqrt{-1})L_{H_i}C^{-1}$ of the ring $\mathcal{H}$. However, $\sigma(C^{-1}) = \sigma(C) = 1$ on $\mathfrak{g}_1^*$ by (6.2) and $\sigma((1/\sqrt{-1})L_{H_i})(e,\xi) = \langle H_i, \xi\rangle$ = pull-back of ξ_i. So, for D_{H_i} the assertion is true. Q.E.D.

Let W be a vector space on which G acts irreducibly. By Lemma 6.1 there exists a subspace, V, of $L^2(G)$ on which $G \times G$ acts irreducibly such that W sits in V as a summand of type (6.1). Let π_V be the orthogonal projection of $L^2(G)$ onto V. Each element, $p(D_H)$, of $\mathcal{H}$ preserves V and preserves the direct sum (6.1). It also preserves each weight space, $W^{(s)}$, of W and on this weight space is equal to

$$p\left(\frac{\beta^{(s)}}{|\beta+\rho|}\right) \text{Identity}$$

$\beta^{(s)} \in \mathfrak{h}^*$ being the weight associated with $W^{(s)}$. Thus

$$\frac{\operatorname{trace} \pi_V\, p(D_H)\, \pi_V}{\dim V} = \frac{1}{\dim W} \sum p\left(\frac{\beta^{(s)}}{|\beta+\rho|}\right) \tag{6.4}$$

the sum taken over all the weights of W, each weight counted with appropriate multiplicity.

Let $W_1, W_2, \cdots$ be a sequence of irreducible representations of G with $N_i = \dim W_i$ tending to infinity. Let $V_1, V_2, \cdots$ be a corresponding sequence of $G \times G$-irreducible subrepresentations of $L^2(G)$. Let β_i be the maximal weight of V_i, and assume $\beta_i/|\beta_i|$ tends to a limit, γ. By Theorem 3

$$\frac{\lim \operatorname{trace} \pi_{V_i}\, p(D_H)\, \pi_{V_i}}{\dim V_i} = \int_{O_\gamma} \sigma(p(D_H))\, d\mu_\gamma \tag{6.5}$$

where O_γ is the $G \times G$-orbit of γ in T^*G and μ_γ the unique $G \times G$-invariant probability measure on O_γ. To evaluate the right hand side of (6.5) we need

LEMMA 6.3. *Let O'_γ be the orbit of γ in $\mathfrak{g}^*$ under the co-adjoint action of G and let μ'_γ be the unique G-invariant probability measure on O'_γ. If f is a continuous G-left-invariant function on $T^*G - 0$ and f' its restriction to $T^*_e = \mathfrak{g}^*$, then*

$$\int_{O_\gamma} f d\mu_\gamma = \int_{O'_\gamma} f' d\mu'_\gamma . \tag{6.6}$$

Proof. The ring of continuous G-left-invariant functions on $T^*G - 0$ is isomorphic with the ring of continuous functions on $T^*_e - 0$, so the left hand side of (6.6) defines *some* measure on $\mathfrak{g}^* - 0$. This measure is obviously a probability measure and is obviously supported on O'_γ. Q.E.D.

Combining (6.4), (6.5), (6.6) and the formula for $\sigma(p(D_H))$ in Lemma 6.2, we get

$$\lim \frac{1}{\dim W_i} \sum p\left(\frac{\beta_i^{(s)}}{|\beta_i + \rho|}\right) = \int_{O'_\gamma} r^* p d\mu'_\gamma = (r_* \mu'_\gamma)(p)$$

for all polynomial functions, p, on $\mathfrak{h}^*$. Since the polynomial functions are dense in the continuous functions, this concludes the proof of Theorem 1.

BIBLIOGRAPHY

[1] L. Boutet de Monvel and V. Guillemin, "Spectral theory for Toeplitz operators," (to appear).

[2] S. Helgason, *Differential Geometry and Symmetric Spaces*, Academic Press, New York (1962).

[3] S. Helgason, "A duality for symmetric spaces with application to group representations," Advances in Math. 5, 1-154 (1970).

[4] J. J. Kohn and L. Nirenberg, "An algebra of pseudodifferential operators," Comm. Pure Appl. Math., 18, 269-305 (1965).

[5] B. Kostant, "On convexity, the Weyl group and the Iwasawa decomposition," Ann. Sci. Ec. Norm. Sup. 6, 413-455 (1973).

[6] G. Schwartz, "Smooth functions invariant under the action of a compact group," Topology, 14, 63-68 (1975).

[7] H. Widom, "Eigenvalue distribution theorems in certain homogeneous spaces," (to appear).

[8] H. Weyl, *The Classical Groups*, Princeton University Press, Princeton, New Jersey (1956).

SOME MICRO-LOCAL ASPECTS OF ANALYSIS ON COMPACT SYMMETRIC SPACES

Victor Guillemin

§1. Let X be a compact Riemannian manifold, Δ the Laplace-Beltrami operator on X and $\gamma \subset X$ a periodic geodesic. It is part of the folklore of spectral theory that the spectrum of Δ contains a sequence of eigenspaces which are in some sense "concentrated" on γ. For instance in [3] and [6] one can find results about "quasi-modes," functions which are asymptotically eigenfunctions of Δ, concentrated on γ. A result of a more precise nature was discovered by us a couple of years ago: Let $X = S^2$, the standard unit sphere in $\mathbf{R}^3$, let $R_\theta : S^2 \to S^2$ be rotation about the z-axis and let $\gamma \subset S^2$ be the geodesic lying in the x, y plane. The eigenvalues of Δ are the integers, $k(k+1)$, $k = 0, 1, 2, \cdots$, the k-th eigenvalue occurring with multiplicity $2k+1$. In the k-th eigenspace there is a normalized eigenfunction, ϕ_k, unique up to constant multiple, with the property that $R_\theta^* \phi_k = e^{ik\theta} \phi_k$ for all $\theta \in (0, 2\pi]$. Let π be the orthogonal projection of $L^2(S^2)$ onto the space spanned by the ϕ_k's. Then π is pseudolocal and has its singular support concentrated on γ; so the ϕ_k's do, in fact, live on γ micro-locally.

In this paper we will extend the result we have just described to an arbitrary compact symmetric space, X. In our generalization, the Laplace operator gets replaced by the whole ring of invariant differential operators

Seminar on Micro-Local Analysis
0-691-08228-6/79/00 0079-33 \$01.65/1 (cloth)
0-691-08232-4/79/00 0079-33 \$01.65/1 (paperback)
For copying information, see copyright page

on X and geodesics by flat, totally geodesic submanifolds. In order to state our result we will need a little notation. Let G be a compact, semi-simple Lie group, let K be a closed subgroup of G and let X be the homogeneous space, G/K. Let $\mathfrak{g}$ and $\mathfrak{k}$ be the Lie algebras of G and K. Then

$$\mathfrak{g} = \mathfrak{k} \oplus \mathfrak{p} \quad \text{(Cartan decomposition)}$$

where $\mathfrak{p}$ is the orthogonal complement of $\mathfrak{k}$ in $\mathfrak{g}$ with respect to the Killing form. X is a symmetric space if $[\mathfrak{p},\mathfrak{p}] \subset \mathfrak{k}$. From now on we will assume this to be the case. Let $\mathfrak{a}$ be a maximal subspace of $\mathfrak{p}$ with the property, $[\mathfrak{a},\mathfrak{a}] = 0$, and let $\mathfrak{h}$ be a Cartan subalgebra of $\mathfrak{g}$ of the form $\mathfrak{h} = \tau \oplus \mathfrak{a}$ with $\tau \subset \mathfrak{k}$. A function, ϕ, on X is called a *conical* function if it is the maximal weight vector in some irreducible sub-representation of $L^2(X)$. (Note that the maximal weight vector in each irreducible sub-representation of $L^2(X)$ is fixed, up to scalar multiple, by the choice of $\mathfrak{h}$.) The irreducible subspaces of $L^2(X)$, and hence the conical functions, are indexed by the points of $\mathfrak{a}_+ \cap \mathfrak{L}^*$, $\mathfrak{a}_+$ being the positive Weyl chamber in $\mathfrak{a}$ and $\mathfrak{L} \subset \mathfrak{a}$ a certain co-compact lattice. We will denote by ϕ_β the conical function indexed by the point, β, in $\mathfrak{a}_+ \cap \mathfrak{L}^*$. Let H^2 be the closed subspace of L^2 spanned by the ϕ_β's, and let π be the orthogonal projection of L^2 onto H^2. This is the obvious analogue of the projection operator described above in the case $X = S^2$. In §3 we show that for each open sub-simplex, S, of the positive Weyl chamber, $\mathfrak{a}_+$, there corresponds a subset, X_S, of the singular support of π. For $S = \text{Int } \mathfrak{a}_+$, X_S is a flat totally geodesic submanifold of X and for $S \subset \partial \mathfrak{a}_+$, X_S is an orbit of the parabolic subgroup, G_S, of G associated with S. The wave-front set of π has an analogous decomposition; to each subsimplex, S, corresponds a component, $\Sigma_S = X_S \times S$, of the wave-front set lying above X_S in the cotangent bundle of X. We will denote by Σ_0 the component of the wave-front set associated with $S = \text{Int } \mathfrak{a}_+$. In §5 we show that on Σ_0, π is pseudolocal, in fact, even a pseudo-differential operator of Hermite type. The proof of this requires results from the paper [2] of Boutet de Monvel and Sjostrand which we briefly review in §4. In §8 we describe, without proof, similar results for the lower dimensional subsimplices of the positive Weyl chamber.

While we were writing this paper, we learned from Joe Wolf about some results of his student, W. Lichtenstein, on the asymptotic properties of ϕ_β, $\beta \in \operatorname{Int} \mathfrak{a}_+$, as $|\beta|$ tends to infinity. We noticed that these results (or at least results very much like them) could be derived easily from the micro-local properties of π described above. We indicate how this is done in §6, and in §7 suggest a way of interpreting these results micro-locally.

This paper would never have got written without the assistance of Sigurdur Helgason, David Vogan and Gregory Zuckerman. I would like to express to them my heart-felt gratitude.

§2. *The defining equations of* H^2

Let $\mathfrak{g}^{\#} = \mathfrak{k} \oplus \sqrt{-1}\,\mathfrak{p}$ be the non-compact form of the Lie algebra, $\mathfrak{g}$. Let $\mathfrak{n}$ be the maximal nilpotent subalgebra of $\mathfrak{g}^{\#}$ in the Iwasawa decomposition, $\mathfrak{g}^{\#} = \mathfrak{k} \oplus \sqrt{-1}\,\mathfrak{a} \oplus \mathfrak{n}$. Let $\mathfrak{m}$ be the centralizer of $\mathfrak{a}$ in k.

Since G acts on X each element, v, of $\mathfrak{g}^{\#}$ determines a (complex-valued) vector field, v', on X. Let

$$D_v : C^\infty(X) \to C^\infty(X) \tag{2.1}$$

be the differential operator, $f \to 1/\sqrt{-1}\, v'f$. Consider the system of equations:

$$D_v f = 0, \quad v \in \mathfrak{n} \tag{2.2}$$

and consider also the system of equations obtained by adding to (2.2) the supplementary equations

$$D_v f = 0, \quad v \in \mathfrak{m}\,. \tag{2.2$'$}$$

PROPOSITION 2.1. *The space of all smooth solutions of* (2.2) *is identical with the space of all smooth solutions of* (2.2) *plus* (2.2)′, *and the* L^2*-closure of this space is* H^2.

Proof. It is enough to check that (2.2) and (2.2) plus (2.2)′ have the same solutions in each irreducible subspace, V, of $L^2(X)$. Since V sits inside of $L^2(X)$, it contains a K-fixed vector by the Frobenius reciprocity theorem; and it is well known (see [12]) that if a space on which G acts irreducibly contains a K-fixed vector, its maximal weight vector is, up to a constant multiple, the unique solution of (2.2), and is also a solution of (2.2)′. Q.E.D.

PROPOSITION 2.2. π *satisfies*

$$D_v \pi = \pi D_v \tag{2.3}$$

for all $v \in \mathfrak{a}$.

Proof. It is enough to check (2.3) on each irreducible subspace, V, of L^2; however (2.3) holds identically on the maximal weight space of V and is zero on the other weight spaces. Q.E.D.

Under the action of $\mathrm{ad}(\mathfrak{a})$, $\mathfrak{n}$ breaks up into a direct sum of one-dimensional subspaces, $\mathfrak{n}_i$, $i = 1, \cdots, N$ where N is the dimension of $\mathfrak{n}$. Moreover, there exists an element, $\alpha_i \in \mathfrak{a}^*$ such that if v_i is a non-zero vector in $\mathfrak{n}_i$

$$[H, v_i] = \sqrt{-1}\,\alpha_i(H)\,v_i \quad \text{for all} \quad H \in \mathfrak{a}\,. \tag{2.4}$$

The α_i's are called the restricted positive roots of G. Let $\mathfrak{a}_+$ be the subset of $\mathfrak{a}$ on which all the α_i's are ≥ 0, and let $(\mathfrak{a}_+)^*$ be the image of $\mathfrak{a}_+$ in $\mathfrak{a}^*$ under the identification of $\mathfrak{a}$ with $\mathfrak{a}^*$ given by the Killing form. Consider the operators

$$A_v = D_v(\Delta + 1)^{-1/2} \tag{2.5}$$

for $v \in \mathfrak{a}$, Δ being the Laplace-Beltrami operator on X. The operators, (2.5), are bounded, self-adjoint and commute with each other.

It is possible to choose a basis $(v_1, \cdots, v_r)$ of $\mathfrak{a}$ with the property that

$$\mathfrak{a}_+ = \left\{\sum c_i v_i, c_i \geq 0\right\}.$$

Let f be a smooth function on $\mathfrak{a}^*$ and let $\tilde{f}$ be the function on R^r obtained by composing f with the mapping

$$R^r \to \mathfrak{a}^*, \qquad (c_1, \cdots, c_r) \to (c_1 v_1 + \cdots + c_r v_r).$$

The spectral theorem allows us to form the operator, $\tilde{f}(A_1, \cdots, A_r)$ as an operator in the uniform closure of the ring of operators generated by $A_1, \cdots, A_r$. Set $f(A) = \tilde{f}(A_1, \cdots, A_r)$.

LEMMA 2.3. $f(A)$ *is a zeroth order pseudodifferential operator. Moreover, its symbol is:*

$$\sigma(f(A)) = \tilde{f}(\sigma(A_1), \cdots, \sigma(A_r)). \tag{2.6}$$

Proof. See [5].

PROPOSITION 2.4. *For all* $f \in C^\infty(\mathfrak{a}^*)$ *with support in the complement of* $(\mathfrak{a}_+)^*$.

$$f(A)\pi = \pi f(A) = 0. \tag{2.2$''$}$$

Proof. Let V be an irreducible subrepresentation of $L^2(X)$, $\sqrt{-1}\,\beta$ the maximal weight of the representation of G on V and v a non-zero vector in the maximal weight space. Then $(\Delta + 1)^{-1/2} v = cv$ with $c > 0$ and $f(A)v = f(c\beta)v$. Since β is in $(\mathfrak{a}_+)^*$, the right hand side of this equation is zero for all f with support in the complement of $\mathfrak{a}^*$.

§3. *The characteristic variety of the defining equations of* H^2

Let x be an arbitrary point of X represented by the coset, gK, of G/K. We recall that the infinitesmal action of $\mathfrak{g}$ on X associates to each $v \in \mathfrak{g}$ a vector field, v', on X. The map

$$\mathfrak{g} \to T_x \tag{3.1}$$

sending v onto $v'(x)$ is surjective. Its kernel is the Lie algebra of the isotropy group at x. Since the isotropy group is gKg^{-1}, the kernel of (3.1) is $(\mathrm{Ad}g)\mathfrak{k}$. The cokernel can, therefore, be canonically identified with $(\mathrm{Ad}g)\mathfrak{p}$. In other words there are canonical identifications

$$T_x^* \cong T_x \cong (\mathrm{Ad}g)\mathfrak{p} \tag{3.2}$$

the identification of T_x^* with T_x being by means of the inner product on T_x coming from the Killing form on $(\mathrm{Ad}g)\mathfrak{p}$. (Note that, since $\mathrm{Ad}K$ maps $\mathfrak{p}$ into itself, $(\mathrm{ad}g)\mathfrak{p}$ depends only on $x = gK$, not on the choice of g.) Given $x = gK$, let

$$\Sigma_x = (\mathrm{Ad}g)\mathfrak{p} \cap \mathfrak{a}_+ - 0 . \tag{3.3}$$

Because of (3.2) we can think of Σ_x both as a subset of T_x^* and as a subset of $\mathfrak{a}_+$. Let Σ be the set-theoretic union of all the Σ_x's. Then we have canonical imbeddings

$$\begin{array}{ccccc} & & T^*X & & \\ & \nearrow & & \searrow & \\ \Sigma & & & & X \\ & \searrow & & \nearrow & \\ & & X \times \mathfrak{a}_+ & & \end{array} \tag{3.4}$$

Let us identify Σ with its image in T^*X.

Proposition 3.1. *Σ is the characteristic variety of the system of equations,* (2.2) *plus* (2.2)′ *plus* (2.2)″.

Proof. Let $\mathfrak{g}^{\#}$ be the non-compact form of $\mathfrak{g}$. It is well known that $\sqrt{-1}\,\mathfrak{a}$ is orthogonal to $\mathfrak{n}$ in the Iwasawa decomposition: $\mathfrak{g}^{\#} = \mathfrak{k} \oplus \sqrt{-1}\,\mathfrak{a} \oplus \mathfrak{n}$; so $\mathfrak{a}$ and $\mathfrak{n}$ are orthogonal in $\mathfrak{g} \otimes \mathbb{C}$. Since $\mathfrak{a}$ is in $\mathfrak{p}$ and $\mathfrak{m}$ is in $\mathfrak{k}$, $\mathfrak{a}$ and $\mathfrak{m}$ are orthogonal in $\mathfrak{g}$. A simple dimension count now shows that the orthogonal complement of $\mathfrak{a}$ in $\mathfrak{g}$ is the space

$$\mathfrak{m} \oplus \mathfrak{g} \cap (\mathfrak{n} \oplus \bar{\mathfrak{n}}) \qquad (\mathfrak{g} \cap \mathfrak{a}^{\perp} \ni \sqrt{-1}(v - \bar{v}) \text{ for all } v \in \mathfrak{n}) .$$

Therefore, by (3.1) and (3.2), the characteristic variety of the system of equations, (2.2) plus (2.2)′, at $x = gK$ is $(\mathrm{Ad}g)\mathfrak{p} \cap \mathfrak{a}$. If we throw in the additional equations, (2.2)″, then by (2.6), this gets cut down to $(\mathrm{Ad}g)\mathfrak{p} \cap \mathfrak{a}_+$. Q.E.D.

The most convenient way of discussing the structure of Σ is as a subset of $X \times \mathfrak{a}_+$. Given $a \in \mathfrak{a}_+ - 0$, let G_a be the centralizer of a in G and let X_a be the orbit of G_a in X containing the identity coset.

PROPOSITION 3.2. *Σ is the union, over all $a \in \mathfrak{a}_+ - 0$ of the sets $X_a \times \{a\}$.*

Proof. Let $x = gK$ and let a_0 be in the intersection of $\mathfrak{a}_+$ with $(\mathrm{Ad}g)\mathfrak{p}$. The map

$$K \times \mathfrak{a}_+ \to \mathfrak{p}$$

sending (k, a) onto $(\mathrm{Ad}\,k)a$ is surjective, (see [7], p. 381); so we can choose g in the coset gK; so that $a_0 \in (\mathrm{Ad}\,g)\mathfrak{a}_+$. Consider now a_0 and $(\mathrm{Ad}\,g)^{-1}a_0$ as elements of the Cartan subalgebra, $\mathfrak{h} = \tau \oplus \mathfrak{a}$, of $\mathfrak{g}$. Since $\mathfrak{a}_+$ is the intersection of $\mathfrak{a}$ with the positive Weyl chamber, $\mathfrak{h}_+$, of $\mathfrak{h}$, both a_0 and $(\mathrm{Ad}\,g)^{-1}a_0$ lie in $\mathfrak{h}_+$. But conjugate elements of $\mathfrak{h}$ lying in the same Weyl chamber are equal, (see, for instance [10], p. 61); so g centralizes a_0. Q.E.D.

Given two elements, a_1 and a_2, of $\mathfrak{a}_+$, we will say that a_1 is equivalent to $a_2 (a_1 \sim a_2)$ if the set of roots which vanish on a_1 is identical with the set of roots which vanish on a_2. The relation, $a_1 \sim a_2$, is easily seen to be an equivalence relation, and its equivalence classes the open subsimplices of $\mathfrak{a}_+$. (We will regard Int $\mathfrak{a}_+$ as being, itself, such a subsimplex.)

LEMMA 3.3. *If* $a_1 \sim a_2$, *then* $G_{a_1} = G_{a_2}$ *and vice versa.*

Proof. See [7], page 249.

If S is an open subsimplex of $\mathfrak{a}_+$, we will define G_S to be the group, G_{a_0}, a_0 being an arbitrary element of S. By the lemma, this definition does not depend on a_0. We will let X_S be the orbit of G_S in X containing the identity coset. From Proposition 3.2, we obtain

PROPOSITION 3.4. Σ *is equal to the disjoint union over all open subsimplices,* S, *of* $\mathfrak{a}_+$, *of the sets,* $\Sigma_S = X_S \times S$, *in* $X \times \mathfrak{a}_+$.

From the results above we can already make some non-trivial conclusions about the projection operator, $\pi: L^2(X) \to H^2(X)$. We recall that if X and Y are manifolds and $A: C_0^\infty(X) \to C^{-\infty}(Y)$ a continuous linear operator, there exists a distributional function, $e_A(x,y)$, on $X \times Y$, called the *Schwartz kernel* of A, such that

$$(Af)(x) = \int_Y e_A(x,y) f(y)\, dy .$$

The *wave-front set* of A is the set of points, (x, ξ, y, η), in $T^*X \times T^*Y$ such that $(x, \xi, y, -\eta)$ is in the wave-front set of e_A. Moreover, A is *regular* if its wave-front set contains no points of the form $(x, \xi, y, 0)$ or $(x, 0, y, \eta)$. If A is regular then it maps $C_0^\infty(X)$ into $C^\infty(Y)$, maps

$C_0^{-\infty}(X)$ into $C^{-\infty}(Y)$, and has a transpose with the same properties. (See [9], §2.4.)

Consider now the operator, π, above. From the results of §2, we can immediately conclude that the wave-front set of π is contained in $\Sigma \times \Sigma$; however, we can conclude a little bit more because of Proposition 2.2. Let $\rho : \Sigma \to \mathfrak{a}_+$ be the projection $(x, a) \to a$ and let Σ^Δ be the fiber product of this mapping with itself.

PROPOSITION 3.5. *The wave-front set of* π *is contained in* Σ^Δ.

Proof. Let (x, ξ, y, η) be in the wave-front set of π. By (3.3), ξ and η are in $\mathfrak{a}_+$; and, by Proposition 2.2, $(\xi, v) = (\eta, v)$ for every $v \in \mathfrak{a}$, $(\ ,\ )$ being the Killing form. Hence $\xi = \eta$. Q.E.D.

COROLLARY. *The operator,* π, *is regular and is a smoothing operator on the set* $X - \cup X_S$.

REMARK. We will see later on that the wave-front set of π is, in fact, considerably smaller than the set, Σ^Δ.

NOTATION. We will denote by Σ_S^Δ the fiber product with itself of $\Sigma_S \to S$. Clearly (3.5) $\Sigma^\Delta = \cup \Sigma_S^\Delta$, union over the subsimplices of $\mathfrak{a}_+$.

§4. *Some facts about Szego kernels*

To obtain more precise micro-local information about π, we will require some general facts about reproducing kernels defined by degenerate elliptic equations. The results we are about to describe are due to Boutet de Monvel and Sjostrand and can be found in §2 of the paper [2].

Let X be a smooth manifold and let $D_1, \cdots, D_N$ be a collection of first order pseudodifferential operators on X. We will say that the system of equations

$$D_1 f = 0, \cdots, D_N f = 0 \tag{4.1}$$

is *in involution* if there exist zeroth order pseudodifferential operators, Q_{ij}^k, such that

$$[D_i, D_j] = \sum Q_{ij}^k D_k$$

for all $1 \leq i, j, k \leq N$. Let σ_i be the symbol of D_i, and let Σ be the set of points in $T^*X - 0$ satisfying: $\sigma_1(x,\xi) = 0, \cdots, \sigma_N(x,\xi) = 0$. At each point (x,ξ) of Σ we define the Levi form of (4.1) to be the $N \times N$ matrix

$$1/\sqrt{-1}\,\{\sigma_i, \bar{\sigma}_j\}(x,\xi)\,. \tag{4.2}$$

The results of Boutet-Sjostrand concern the micro-local behavior of the system (4.1) at points where the Levi form is positive definite. It is a simple exercise in symplectic geometry to show that Σ is a manifold of codimension 2N at such points. It is also easy to see that Σ is symplectic at such points: the restriction to Σ of the symplectic form on T^*X is non-degenerate.

To describe the results of [2] we need to introduce a little terminology. Let X be a differentiable manifold and A_1 and A_2 continuous linear operators on $C^{-\infty}(X)$. Given an open conic subset, $\mathcal{C}$, of $T^*X - 0$ we will say that A_1 and A_2 *are equivalent on* $\mathcal{C}$ ($A_1 \sim A_2$ on $\mathcal{C}$) if for every generalized function, u, with wave-front set contained in $\mathcal{C}$ $(A_1 - A_2)u$ is smooth.

THEOREM 4.1. *Let* (x,ξ) *be a point in* Σ *at which the Levi form* (4.2) *is positive definite. Then there exists a conic neighborhood,* $\mathcal{C}$, *of* (x,ξ) *and regular operators,* $\pi, L_1, \cdots, L_N$ *such that*

(i) $\pi \sim \pi^t \sim \pi^2$ *on* $\mathcal{C}$

(ii) $I \sim \pi + \sum L_i D_i$ *on* $\mathcal{C}$

(iii) $D_i \pi \sim 0$ *on* $\mathcal{C}$ *for all* i.

Moreover, π *is uniquely determined up to equivalence by these properties.*

Proof. See [2], Theorem 2.14.

A companion theorem to Theorem 4.1 describes more precisely the nature of the operator, π. Suppose, in general, we are given a differentiable manifold, X and a conic symplectic submanifold, Σ, of $T^*X - 0$. In [1] Boutet associates with the pair (X, Σ) a natural class of symbols $\mathcal{H}^k(X, \Sigma)$ which are of type $S^k_{1/2, 1/2}$ on all of $T^*X - 0$ and are of type $S^{-\infty}$ outside every conic neighborhood of Σ. These symbols are called *Hermite* symbols and the space of operators associated with them: $OP\mathcal{H}^k(X,\Sigma)$, are called *Hermite operators*. Boutet proves in [1] the following facts about these operators.

1. $A \in OP\mathcal{H} \Rightarrow WF(A)$ is contained in the diagonal in $\Sigma \times \Sigma$.
2. If $A \in OP\mathcal{H}^k$ and Q is an ordinary pseudodifferential operator of order ℓ then AQ and QA are contained in $OP\mathcal{H}^{k+\ell}$.
3. If $A \in OP\mathcal{H}^k$ and $B \in OP\mathcal{H}^\ell$ then $AB \in OP\mathcal{H}^{k+\ell}$.
4. If $A \in OP\mathcal{H}^k$, $A^t \in OP\mathcal{H}^k$.

In [4] it is shown that Hermite operators have intrinsically defined leading symbols. The theory of these symbols involves the metaplectic group and the "symplectic spinors" of Kostant. An example of a Hermite operator which "occurs in nature" is the following. Let X be a strictly pseudo-convex domain in $\mathbb{C}^n$ and let H^2 be a L^2-closure in $L^2(\partial X)$ of the space of C^∞ functions on ∂X which are restrictions of holomorphic functions on X. Associated with the contact structure on ∂X is a conic symplectic submanifold, Σ, of $T^*\partial X - 0$. In [2] Boutet and Sjostrand show that the Szego projector, $\pi : L^2(\partial X) \to H^2(\partial X)$, is a Hermite operator; in fact they show that π belongs to $OP\mathcal{H}^0(\partial X, \Sigma)$. More generally if $q \in C^\infty(\partial X)$ the Toeplitz operator

$$f \in H^2 \to \pi qf$$

belongs to $OP\mathcal{H}^0(\partial X, \Sigma)$. Getting back to the system of equations (4.1) Boutet and Sjostrand prove

THEOREM 4.2. *The projector* π *described in Theorem 4.1 is in* $OP\mathcal{H}^0(X, \Sigma)$.

§5. *Micro-local properties of* π *in the interior of the positive Weyl chamber*

We recall that the nilpotent algebra, $\mathfrak{n}$, can be decomposed into a direct sum of one-dimensional pieces

$$\mathfrak{n} = \sum \mathfrak{n}_i$$

such that $\mathfrak{n}_i$ is Ad($\mathfrak{a}$)-invariant and

$$[H, v_i] = \sqrt{-1}\, \alpha_i(H)\, v_i \quad \text{for all} \quad H \in \mathfrak{a} \tag{5.1}$$

v_i being any non-zero vector in $\mathfrak{n}_i$. (See (2.4).) To each $\alpha_i \in \mathfrak{a}^*$ there corresponds a unique vector $H_i \in \mathfrak{a}$ such that for all $H \in \mathfrak{a}$, $\alpha_i(H) = (H_i, H)$, $-(\ ,\ )$ being the Killing form.

LEMMA 5.1. *If* v_i *is a non-zero vector in* $\mathfrak{n}_i$, *then* $1/\sqrt{-1}\,[v_i, \overline{v}_i] = c_i H_i$ *with* $c_i > 0$.

Proof. See [7], page 246.

COROLLARY. *One can choose a basis* $v_1, \cdots, v_N$ *of* $\mathfrak{n}$ *such that* (5.1) *holds and, in addition,*

$$1/\sqrt{-1}\,[v_i, \overline{v}_i] = H_i \tag{5.2}$$

where $(H_i, H) = \alpha_i(H)$ *for all* $H \in \mathfrak{a}$.

Let P be the differential operator:

$$P = \sum_{i=1}^{N} D_{v_i}^t D_{v_i}\,.$$

PROPOSITION 5.2. *There exists a constant,* $C > 0$, *such that if* $f \in C^\infty(X)$ *is perpendicular to the kernel of* π, *then* $(Pf, f) \geq C\,\|f\|_2^2$.

Proof. By Lemma 5.1, $v_i, \overline{v}_i$ and $\sqrt{-1}\,H_i$ satisfy the bracket relations

$$[v_i, \sqrt{-1}\,H_i] = a_i v_i\,,\quad [\overline{v}_i, \sqrt{-1}\,H_i] = -a_i \overline{v}_i\,,\quad [v_i, \overline{v}_i] = \sqrt{-1}\,H_i\,,$$

where $a_i = (H_i, H_i) > 0$. Set $v^+ = 1/\sqrt{a_i}\,V_i$, $v^- = 1/\sqrt{a_i}\,\overline{v}_i$ and $z = 1/a_i\,\sqrt{-1}\,H$. Then v^+, v^- and z satisfy the bracket relations

$$[v^+, z] = v^+\,,\quad [v^-, z] = -v^-\,,\quad [v^+, v^-] = z\,; \tag{5.3}$$

i.e. v^+, v^- and z are the standard basis for $sl(2, \mathbf{R})$. For the following lemma, see [11].

LEMMA 5.3. *Let* $V^{(m)}$ *be an* $m{+}1$*-dimensional irreducible* $sl(2, \mathbf{R})$ *module. Then*

$$V^{(m)} = \sum_{i=0}^{m} V_i^{(m)}$$

where z *is equal to* $(m-2i)$ *times the identity on* $V_i^{(m)}$ *and* v^-v^+ *is equal to* $i(m-i+1)I$ *on* $V_i^{(m)}$.

Let V be any irreducible subspace of $L^2(X)$. Under the action of $ad(\mathfrak{a})$, V decomposes into a direct sum of one-dimensional weight spaces

$$V_{max} \oplus \sum V_\beta\,,$$

V_{max} being the maximal weight space. If f is a non-zero element of V_β, then for some v_i, $D_{v_i} f \neq 0$. Applying the lemma to the rescaled vectors (5.3) we get

$$(D_{v_i}^t D_{v_i} f, f) = \|D_{v_i} f\|_2^2 \geq a_i \|f\|_2^2\,.$$

For the rest of the v_j's we have

$$(D_{v_j}^t D_{v_j} f, f) = \|D_{v_j} f\|_2^2 \geq 0\,;$$

so with $C = \min a_i$ we get

$$(Pf, f) \geq C\|f\|^2$$

for all f in the orthogonal complement of V_{max} in V. Since $L^2(X)$ is a direct sum of its irreducible subspaces this proves Proposition 5.2.

Q.E.D.

Let Δ be the standard Laplace-Beltrami operator on X. For every irreducible subspace, V of $L^2(X)$, Δ maps V into V and, in fact, is equal to a constant multiple of the identity on V. Hence Δ commutes with P. Let Q be the unique operator which is equal to zero on the range of π and equal to P^{-1} on the Kernel of π. By Proposition 5.2, Q is a bounded operator from L^2 to L^2.

PROPOSITION 5.4. *Q is a bounded operator from H^s to H^s for all s and is regular in the sense of §3.*

Proof. Since P commutes with Δ, so does Q. However $(\Delta+1)^{s/2}$ maps L^2 isomorphically onto H^s; so Q is a bounded operator from H^s to H^s. Suppose (x, ξ, y, η) is in the wave-front set of Q. Since $\Delta Q = Q\Delta$, $\sigma(\Delta)(x,\xi) = \sigma(\Delta)(y,\eta)$, i.e. $|\xi| = |\eta|$. In particular if $\xi = 0$, $\eta = 0$ and vice-versa. Q.E.D.

Define the operator

$$L_i : C^\infty(X) \to C^\infty(X)$$

to be the operator: $L_i f = QD^t_{v_i} f$. Then L_i is regular and satisfies

$$\text{Identity} = \pi + \sum L_i D_{v_i} . \tag{5.4}$$

In addition, of course, π satisfies

$$\pi = \pi^2 = \pi^t \tag{5.5}$$

and

(5.6) $$D_{v_i}\pi = 0 \quad \text{for} \quad i = 1, \cdots, N .$$

Therefore, π satisfies the conditions (i), (ii) and (iii) of Theorem (4.1); (in fact, it satisfies these conditions globally not just micro-locally).

We will now use the uniqueness assertion in Theorem 4.1 to draw some conclusions about the structure of π in the interior of the positive Weyl chamber. First, however, we will need some notation. Let $S_o =$ Interior $\mathfrak{a}_+$, let $G_o = G_{S_o}$ and let $X_o = X_{S_o}$. Let A be the connected Lie subgroup of G having $\mathfrak{a}$ for its Lie algebra. The group, A, is compact ([7], p. 210) and abelian. Let M be its centralizer in K.

LEMMA 5.5. $G_o = AM$ *and* X_o *is the orbit of* A *in* X *containing the identity coset.*

Proof. For the first assertion, see [7], Chapter VII.

If o is the point in X representing the identity coset, then M is contained in the isotropy group K, of o; so $X_o = AM \cdot o = A \cdot o$. Q.E.D.

The group $A \cap K$ is abelian and is also discrete since $\mathfrak{a} \cap \mathfrak{k} = \{0\}$. Therefore

$$X_o = A/A \cap K$$

is a compact abelian group of the same dimension as A. One can show that X_o, viewed as a submanifold of X, is flat and totally geodesic. Moreover, every flat, totally geodesic submanifold of X of the same dimension as A is of the form, gX_o, for some $g \in G$. (For the proof of these assertions see [7], page 210.) Consider now the open subset, $\Sigma_o = X_o \times S_o$, of Σ. We will prove:

THEOREM 5.6. *If* $(x, \xi) \in \Sigma$ *the system of equations*

(5.7) $$D_{v_i} f = 0 , \quad i = 1, \cdots, N$$

has positive definite Levi form at (x, ξ).

Proof. The Levi form at (x,ξ) is the $N \times N$ matrix

$$1/\sqrt{-1}\,(\xi,[v_i,\overline{v}_j]), \quad 1 \le i,j \le N. \tag{5.8}$$

If $i \neq j$, $[v_i,\overline{v}_j]$ is in $\mathfrak{n} \oplus \overline{\mathfrak{n}}$; so, with ξ in $\mathfrak{a}$, (5.8) is zero. If $i = j$, then (5.8) becomes just (ξ, H_i) by Lemma 5.1. Since $\xi \in \operatorname{Int} \mathfrak{a}_+$, (5.8) is positive when $i = j$. This shows that (5.8) is a diagonal matrix with positive entries along the diagonal. Q.E.D.

COROLLARY 1. *The characteristic variety of the system of equations (5.7) contains Σ_o as an open subset.*

Proof. Since the Levi form is non-degenerate at (x,ξ) the characteristic variety of the system (5.7) is locally a submanifold of dimension equal to twice the dimension of X minus twice the dimension of $\mathfrak{n}$. It is easy to check that this is also the dimension of Σ_o.

COROLLARY 2. *Let (x,ξ) and (y,η) be elements of Σ. If $(x,\xi,y,\eta) \in WF(\pi)$ and either (x,ξ) or (y,η) is in Σ_o, then $(x,\xi) = (y,\eta)$.*

Proof. Suppose $(x,\xi) \in \Sigma_o$. By Proposition 3.5, $\xi = \eta$, so (y,η) is also in Σ_o. It remains to show that $x = y$. If this were not the case, then by Proposition 4.1 we could find a conic neighborhood, $\mathcal{C}_1$, of (x,ξ) not containing (y,η) and operators, π', $L'_1, \cdots, L'_N$ such that

(i) $\pi' \sim (\pi')^2 \sim (\pi')^t$

(ii) $I \sim \pi' + \sum L'_i D_{v_i}$

and

(iii) $D\pi'_i \sim 0$, $i = 1, \cdots, N$

on $\mathcal{C}_1$. Let $\mathcal{C}_2$ be a conic neighborhood of (y,η) not intersecting $\mathcal{C}_1$. By (5.4), $\pi \sim -\Sigma D^t_{v_i} L_i$ on $\mathcal{C}_1 \times \mathcal{C}_2$. In addition, $\pi' D^t_{v_i} \sim (D_{v_i}\pi')^t \sim 0$

on $\mathcal{C}_1$. Therefore, we conclude

$$\pi \sim (\pi' + \Sigma L'_i D_{v_i})\pi \sim \pi'\pi \sim \pi'(-\Sigma D^t_{v_i} L_i) \sim 0$$

on $\mathcal{C}_1 \times \mathcal{C}_2$. This proves that π is smoothing on $\mathcal{C}_1 \times \mathcal{C}_2$ which is a contradiction. Q.E.D.

COROLLARY 3. *On* Σ_o, π *is micro-locally in* $OP\mathcal{H}^0(X, \Sigma_o)$.

We will next show that we can modify π a little on the boundaries of the Weyl chamber so as to get an operator which is *globally* in $OP\mathcal{H}^0(X, \Sigma_o)$. To start with, we will use the Killing form to identity $\mathfrak{a}_+$ with $\mathfrak{a}_+^*$. Let f be a function on $\mathfrak{a}_+$ which is smooth except at the origin, homogeneous of degree zero, and vanishes on a neighborhood of $\partial\mathfrak{a}_+ - 0$. If V_β is an irreducible subspace of $L^2(X)$ and β the maximal weight of the representation of G on V_β, then $\beta \in \mathfrak{a}_+$. Define an operator $\pi_f : L^2(X) \to L^2(X)$ as follows:

1) If V_β is a non-zero maximal weight vector in $V = V_\beta$ set $\pi_f(V_\beta) = f(\beta)V_\beta$.
2) If V is orthogonal to the maximal weight vector in V set $\pi_f(V) = 0$.

Together, 1) and 2) define π_f unambiguously on all of $L^2(X)$. This operator is a kind of smoothed out version of π in which the pathologies occurring at the boundary of $\mathfrak{a}_+$ have been eliminated.

THEOREM 5.7. π_f *belongs to* $OP\mathcal{H}^0(X, \Sigma_o)$.

Proof. Let f(A) be the operator defined at the end of section 2. (See Lemma 2.3.) By Lemma 2.3, f(A) is an ordinary pseudodifferential operator of order zero. We will show that the wave-front set of $f(A)\pi$ is concentrated on Σ_o. To see this, let g be any homogeneous function order zero on $\mathfrak{a}$ which is smooth except at the origin. If f and g have non-overlapping supports

$$f(A)\,g(A) = 0 \,. \tag{5.9}$$

Let (x,ξ) be a point of $\Sigma - \Sigma_o$. We can choose g such that $g(\xi) \neq 0$ and g and f have non-overlapping supports. By 2.6 g is elliptic at (x,ξ); so $f(A)$ is smoothing there. Therefore, the wave-front set of $f(A)\pi$ is concentrated on the diagonal in $\Sigma_o \times \Sigma_o$. From Corollary 3 of Theorem 5.6 we conclude that $f(A)\pi \in OP\mathcal{H}^0(X,\Sigma_o)$. We will now show that $\pi_f = f(A)\pi$. To see this let V_β be an irreducible subrepresentation of $L^2(X)$ with maximal weight β and maximal weight vector v_β. Let $(\Delta+1)^{-1} v_\beta = c v_\beta$. Then, by definition of $f(A)$,

$$f(A)\pi v_\beta = f(A)v_\beta = f(c\beta)v_\beta \,.$$

Since f is homogeneous of degree zero the term on the right is $f(\beta)v_\beta$ or $\pi_f v_\beta$. Both π_f and $f(A)\pi$ are zero on the orthogonal complement of v_β in V_β; so they are identical on V_β. V_β being arbitrary, they must be identical on $L^2(X)$.

§6. *Some asymptotic properties of conical functions on the interior of the positive Weyl chamber*

Let us introduce the following notation. If V is an irreducible subrepresentation of $L^2(X)$ with maximal weight, β, let v_β be its maximal weight vector. We will normalize v_β so that its L^2 norm (as an element of $L^2(X)$) is one. It is well known that every irreducible representation of G which occurs as a subrepresentation of $L^2(X)$ occurs with multiplicity one; so v_β is determined completely by β up to multiplicative constant of modulus one. To distinguish between v_β regarded as a vector and v_β regarded as a function on X we will denote the function which v_β represents by ϕ_β. ϕ_β's are the *conical functions* on X.

Let $\mathcal{W}$ be a conic subset of Int $\mathfrak{a}_+$ whose closure in $\mathfrak{a} - 0$ is properly contained in Int $\mathfrak{a}_+$. We will show that the conical functions have the following asymptotic properties. (The first two of these properties were discovered by W. Lichtenstein.)

PROPOSITION 6.1. *Let* q *be in* $C_0^\infty(X - X_o)$. *Then*

$$\int_X q|\phi_\beta|^2 x = O(|\beta|^{-k}) \tag{6.1}$$

for all $\beta \in \mathfrak{W}$ *and* $k > 0$.

PROPOSITION 6.2. *Let* q *be in* $C^\infty(X)$. *Then*

$$\int_X q|\phi_\beta|^2\, dx = \int_{X_o} q\, dx_o + O(|\beta|^{-1}) \tag{6.2}$$

for all $\beta \in \mathfrak{W}$, dx_o *being normalized Haar measure.*

COROLLARY. *The function,* $|\phi_\beta|^2$ *converges to the delta function,* δ_{X_o}, *as* β *tends to infinity in* $\mathfrak{W}$.

Proposition 6.2 can be generalized:

PROPOSITION 6.3. *Let* $Q: L^2(X) \to L^2(X)$ *be a zeroth order pseudo-differential operator with leading symbol,* $\sigma(Q)$. *Then for all* $\beta \in \mathfrak{W}$

$$\int_X Q\phi_\beta\bar{\phi}_\beta(x)\, dx = \int_{X_o} \sigma(Q)(x, \beta)\, dx + O(|\beta|^{-1})\,. \tag{6.3}$$

We will first prove Proposition 6.1. Let f be a smooth function on $\mathfrak{a}_+ - 0$ which is homogeneous of degree zero, supported in Int $\mathfrak{a}_+$ and equal to one on $\mathfrak{W}$. Let π_f be the operator described in Theorem 5.7.

Let q be a smooth function on X with support in $X - X_o$ and let $M_q : L^2(X) \to L^2(X)$ be the operator "multiplication by q." By Theorem 5.7, $M_q \pi_f$ is smoothing; so, for all k, $M_q \pi \Delta^k$ is smoothing and, in particular, bounded as an operator from L^2 to L^2. Let $\rho \in \mathfrak{a}_+$ be half the sum of the restricted roots. Then, by [7],

$$\Delta \phi_\beta = (|\beta+\rho|^2 - |\rho|^2)\phi_\beta \tag{6.4}$$

for all β. Since $\pi_f \phi_\beta = \phi_\beta$ when $\beta \in \mathfrak{W}$ we conclude from the L^2-boundedness of $M_q \pi \Delta^k$:

$$C \geq \langle M_q \pi_f \Delta^k \phi_\beta, \phi_\beta \rangle_{L^2} = (|\beta+\rho|^2 - |\rho|^2)^k \int_X q|\phi_\beta|^2 \, dx$$

when $\beta \in \mathfrak{W}$. This proves Proposition 6.1.

To prove Proposition 6.2 we first note that if $a \in A$, $q \in C^\infty(X)$, and q_a is the translate of q by a; i.e. $q_a(x) = q(ax)$, then for all β,

$$\langle q\phi_\beta, \phi_\beta \rangle_{L^2} = \langle q_a \phi_\beta, \phi_\beta \rangle_{L^2}$$

since $|\phi_\beta|^2(ax) = |\phi_\beta|^2(x)$. Therefore, if we set

$$q_{av.}(x) = \int_A q_a(x)\, da ,$$

da being normalized Haar measure on A,

$$\langle q\phi_\beta, \phi_\beta \rangle_{L^2} = \langle q_{av.} \phi_\beta, \phi_\beta \rangle_{L^2} .$$

Let $c = \int_{X_o} q \, dx_o$. Then $q_{av.}$ is equal to c on X_o; so if we set $q_o = q_{av.} - c$, we have

$$\langle q\phi_\beta, \phi_\beta \rangle_{L^2} = c + \langle q_o \phi_\beta, \phi_\beta \rangle_{L^2} \tag{6.5}$$

with q_o vanishing on X_o. Now $\pi M_{q_o} \pi_f$ is a zeroth order Hermite operator whose leading symbol vanishes, since q_o vanishes on X_o; so it can be written in the form $\pi Q \pi$ where Q is a standard pseudodifferential operator of order -1. The second term in (6.5) can be rewritten:

$$(6.6) \qquad (|\beta+\rho|^2 - |\rho|^2)^{-\frac{1}{2}} < Q\Delta^{\frac{1}{2}}\phi_\beta, \phi_\beta >_{L^2} .$$

The second factor is bounded in β; so (6.6) is of order $0(|\beta|^{-1})$. This proves Proposition 6.2.

The proof of Proposition 6.3 is similar. Let us denote by $T_a : X \to X$ the mapping $x \to ax$. If Q is a pseudodifferential operator of order zero,

$$< Q\phi_\beta, \phi_\beta >_{L^2} = < T_a^* Q T_{-a}^* \phi_\beta, \phi_\beta >_{L^2} ,$$

so it is enough to prove (6.3) for the operator,

$$\int_A T_a^* Q T_{-a}^* \, da ;$$

i.e. it is enough to prove the theorem for operators which commute with T_a for all $a \in A$. However, every such operator can be written in the form:

$$f(A) + Q_o$$

where f is a smooth homogeneous function of degree zero on $\mathfrak{a} - 0$, $f(A)$ is defined as at the end of section 2, and Q_o has a symbol which vanishes on Σ_o. Now $< Q_o\phi_\beta, \phi_\beta >_{L^2} = 0(|\beta|^{-1})$ by the same reasoning as above, and

$$(6.6) \qquad < f(A)\phi_\beta, \phi_\beta >_{L^2} = f(\beta) .$$

Combining (6.6) with Lemma 2.3, we get (6.3).

§7. *The restrictions of conical functions to* X_o

We recall that X_o is the orbit of A containing the identity coset, $o \in X$. Therefore, there is a mapping:

$$\kappa : \mathfrak{a} \to X_o \tag{7.1}$$

mapping $a \in \mathfrak{a}$ onto $(\exp a)\cdot o$. The preimage, $\kappa^{-1}(o)$, consists of all $a \in \mathfrak{a}$ such that $\exp a \in K$. Let us denote this preimage by $\mathfrak{L}$. It is easy to see that $\mathfrak{L}$ is a co-compact lattice in $\mathfrak{a}$. Given a conical function, ϕ_β, the restriction of ϕ_β to X_o transforms under the action of A like the group character $e^{\sqrt{-1}\,\beta(\log a)}$; so, either ϕ_β, restricted to X_o must vanish, or β must be in the dual lattice, $\mathfrak{L}^*$. It turns out that only the second alternative can occur. For the following, see [8],

PROPOSITION 7.1. *The conical functions are in* 1-1 *correspondence with the points in* $\mathfrak{L}^* \cap \mathfrak{a}_+$. *More specifically, for each* $\beta \in \mathfrak{L}^* \cap \mathfrak{a}_+$, *there exists a conical function,* ϕ_β, *such that*

$$\phi_\beta \mid X_o = c_\beta e^{i(\beta, x)} \tag{7.2}$$

with $c_\beta \neq 0$.

We will investigate the restriction mapping, (7.2), from the micro-local point of view. Let f be a smooth function on $\mathfrak{a}_+ - 0$ which is homogeneous of degree zero and is supported in $\operatorname{Int} \mathfrak{a}_+$. In §5 we associated with f the operator $\pi_f \in OP\mathcal{H}^0(X, \Sigma_o)$. We will now show that this is closely related to the operator

$$P_f : L^2(X_o) \to L^2(X_o) \tag{7.3}$$

defined by setting $P_f e_\beta = f(\beta) e_\beta$ for all $\beta \in \mathfrak{L}^*$. (Here we have set $e_\beta(x) = e^{\sqrt{-1}\,(\beta, x)}$.) Note that P_f is a pseudodifferential operator of order zero and that it commutes with the action of A on X_o.

LEMMA 7.2. *Let* i_{X_o} *be the inclusion mapping of* X_o *into* X. *Then*

$$i^*_{X_o}\pi_f = P_f i^*_{X_o}\pi . \tag{7.4}$$

Proof. This is an immediate consequence of (7.2).

A simple argument, based on wave-front set considerations, shows that $i^*_{X_o}\pi$ is regular; so its transpose is well-defined as an operator from $C^\infty(X_o)$ to $C^\infty(X)$. We will denote this transpose by W and the transpose of $i^*_{X_o}\pi_f$ by W_f. From (7.4) we get

$$W_f = WP_f . \tag{7.5}$$

We will show in a moment that (7.5) can be considerably generalized.

LEMMA 7.3. W^tW_f *is a pseudodifferential operator of order* $-N$, *where* $N = \dim X - \dim X_o$. *The symbol of this operator is non-vanishing wherever the symbol of* P_f *is non-vanishing.*

The proof of this lemma involves standard composition formulas for Hermite operators, for which we refer to [1].

REMARKS:

1) The operator, W^tW_f, commutes with the action of A on X_o, so it is a convolution operator.

2) Though we won't need this in what follows, an explicit formula can be given for the total symbol of W^tW_f, namely

$$\sigma_{\text{total}}(W^tW_f)(\beta) = \frac{c(\beta+\rho)}{w(\beta)^2} f(\beta) \tag{7.6}$$

where c is the Harish-Chandra c-function and w is the polynomial occurring on the right hand side of the Weyl character formula.

By Lemma 7.3 we can find an invertible pseudodifferential operator, C, on X_o which is positive, self-adjoint, elliptic of order $-N/2$, A-invariant and satisfies

$$W^t W_f = C^2 P_f = C P_f C . \tag{7.7}$$

We will define

$$U = WC^{-1} \quad \text{and} \quad U_f = W_f C^{-1} . \tag{7.8}$$

Then, by definition,

$$U^t U_f = P_f . \tag{7.9}$$

Let $di_{X_o} : TX_o \to TX$ be the derivative of the inclusion mapping of X_o into X. The Reimannian inner products on X and X_o give us identifications

$$TX \cong T^*X \quad \text{and} \quad TX_o \cong T^*X_o ;$$

so, from di_{X_o} we get a mapping

$$j : T^*X_o \to T^*X . \tag{7.10}$$

This mapping is symplectic in the weak sense that the pull-back of the canonical one-form on T^*X is the canonical one form on T^*X_o.

PROPOSITION 7.4. *U_f is a Hermite-Fourier Integral operator of order zero, associated with the symplectic mapping* (7.10).

At the end of §4, we attempted to explain, in terms the ordinary layman can understand, the theory of Hermite pseudodifferential operators. For the theory of Hermite Fourier Integral Operators and their composition properties, of which the above result is an immediate consequence, we refer to [1].

Let $\mathcal{O}$ be a convex, conic subset of $\mathfrak{a}_+$ on which $f \equiv 1$. Then, for

$\beta \in \mathfrak{W}$, $P_f e_\beta = e_\beta$; so U_f maps the subspace spanned by e_β, $\beta \in \mathfrak{W}$, isometrically into $H^2(X)$. (Recall that $e_\beta(x) = e^{\sqrt{-1}(\beta,x)}$.) From this and the A-invariance of U_f we conclude:

LEMMA 7.5. *For all* $\beta \in \mathfrak{W}$, $U_f e_\beta = \phi_\beta$.

The following is a sharpening of Proposition 6.3.

PROPOSITION 7.6. *Let* Q *be a pseudodifferential operator on* X. *Then there exists a pseudodifferential operator,* Q_o, *on* X_o *such that*

$$(\pi Q \pi) U_f = U_f Q_o . \tag{7.11}$$

Moreover, on the subset of T^*X_o *where* $\sigma(P_f) \neq 0$, *the symbol of* Q *and the symbol of* Q_o *are related by:*

$$\sigma(Q_o) = j^*\sigma(Q) , \tag{7.12}$$

j *being the mapping* (7.10).

Proof. Let f_1 be a smooth function, homogeneous of degree zero on $\mathfrak{a}_+ - 0$ with support in Int $\mathfrak{a}_+$ and with the property $f_1 \equiv 1$ on support f. Set

$$Q_o = U^t Q U_{f_1} .$$

It is easy to check that this has the desired properties.

COROLLARY. *If* β *and* β' *are in* $\mathfrak{W}$ *then*

$$\int_X Q\phi_\beta \bar{\phi}_{\beta'} dx = \int_{X_o} Q_o e_\beta \bar{e}_{\beta'} dx_o . \tag{7.13}$$

Proof. The integral on the left is $<\pi Q \pi \phi_\beta, \phi_{\beta'}>_{L^2}$. If β and β' are in $\mathfrak{W}$, we have

$$
\begin{aligned}
<\pi Q \pi \phi_\beta, \phi_{\beta'}> &= <\pi Q \pi U_f e_\beta, U_f e_{\beta'}> \\
&= <U_f Q_o e_\beta, U_f e_{\beta'}> \\
&= <Q_o e_\beta, U_f^t U_f e_{\beta'}> \\
&= <Q_o e_\beta, e_{\beta'}> ,
\end{aligned}
$$

which is the integral on the right in 7.13.

§8. *Micro-local properties of* π *on the walls of the positive Weyl chamber*

We will describe briefly how π behaves on the other pieces of its characteristic variety, namely on the sets, Σ_S, S ranging over the subsimplices of $\partial \mathfrak{a}_+$. To begin with we will introduce a partial ordering among the subsimplices of $\mathfrak{a}^+$ by setting $S < T$ if $S \subset \partial T$. Given a subsimplex, S, of $\mathfrak{a}_+$, let H_S^2 be the L^2 closure of the space spanned by the conical functions, ϕ_β, β an element of the closure of S in $\mathfrak{a}_+$. Let π^S be the orthogonal projection of L^2 onto H_S^2.

THEOREM 8.1. *The wave-front set of* π^S *is contained in the union, over all* $S' \leq S$, *of the sets* $\Sigma_{S'}^\Delta$. *Moreover* $\pi \sim \pi^S$ *on* $\Sigma^\Delta - \bigcup_{S' > S} \Sigma_{S'}^\Delta$.

The proof of this will be given elsewhere. (For the definition of Σ_S^Δ, see (3.5).)

We recall that if a_1 and $a_2 \in S$ then the centralizer, G_{a_1}, of a_1 in G is equal to the centralizer, G_{a_2}, of a_2 in G. In §3 we defined G_S to be equal to G_a, a being a representative element of S. Let $\mathfrak{g}_S$ be the Lie algebra of G_S and let $\mathfrak{h}_S$ be the following subspace of $\mathfrak{g}_S$.

(8.1) $$\mathfrak{h}_S = \{v \in \mathfrak{g}, [v,a] = 0, (v,a) = 0, \forall a \in S\} .$$

LEMMA 8.2. *$\mathfrak{h}_S$ is a Lie subalgebra of $\mathfrak{g}_S$.*

Proof. Given elements v_1 and v_2 of $\mathfrak{h}_S$, $[[v_1, v_2], a] = 0$ for all $a \in S$ by Jacobi's identity. Also we have:

$$\begin{aligned}([v_1, v_2], a) &= \mathrm{tr}(\mathrm{ad}([v_1, v_2])\, \mathrm{ad}a) \\ &= \mathrm{tr}(\mathrm{ad}v_1\, \mathrm{ad}v_2\, \mathrm{ad}a) - \mathrm{tr}(\mathrm{ad}v_2\, \mathrm{ad}v_1\, \mathrm{ad}a) \\ &= \mathrm{tr}(\mathrm{ad}v_1\, \mathrm{ad}a\, \mathrm{ad}v_2) - \mathrm{tr}(\mathrm{ad}v_2\, \mathrm{ad}v_1\, \mathrm{ad}a) \\ &= 0\,.\end{aligned}$$

(In the second-to-last line we used the fact that $\mathrm{ad}v_2$ and $\mathrm{ad}a$ commute.)

Q.E.D.

Let H_S be the connected Lie subgroup of G_S associated with $\mathfrak{h}_S$. It is not hard to see that H_S is closed in G_S. In addition we have:

LEMMA 8.3. *H_S is a normal subgroup of G_S and the quotient group, G_S/H_S, is abelian.*

Proof. Let $\mathfrak{a}_S$ be the abelian subalgebra of $\mathfrak{a}$ spanned by $\{a, a \in S\}$. It is clear from (8.1) that

$$\mathfrak{g}_S = \mathfrak{a}_S \oplus \mathfrak{h}_S\,. \tag{8.2}$$

Since $\mathfrak{h}_S$ is in the centralizer of $\mathfrak{a}_S$, it is a subideal of $\mathfrak{g}_S$. This shows that H_S is a normal subgroup of G_S. By (8.2) the Lie algebra of G_S/H_S is $\mathfrak{a}_S$; so G_S/H_S is abelian. Note that G, hence G_S is connected.

G.E.D.

PROPOSITION 8.4. *If β is in the closure of S in $\mathfrak{a}_+$, the conical function, ϕ_β, is H_S-invariant.*

Proof. We decompose $\mathfrak{n}$ into one-dimensional root spaces, $\mathfrak{n}_i$, associated with the roots, α_i, and choose $v_i \in \mathfrak{n}_i$, $\bar{v}_i \in \bar{\mathfrak{n}}_i$ and $H_i = 1/\sqrt{-1}\,[v_i, \bar{v}_i]$ as in the corollary to Lemma 5.1. Let us number the roots, α_i, $i = 1, \cdots, N$ so that the first N' roots are precisely the roots which vanish on S. It is clear that $\mathfrak{h}_S \otimes \mathbb{C}$ is spanned by $\mathfrak{m}$ and by the vectors $v_i, \bar{v}_i$ and H_i, $1 \leq i \leq N'$. Let V be the irreducible subspace of $L^2(X)$ in which ϕ_β sits and let v_β be the maximal weight vector in V: the abstract element in V which is represented concretely on X by the function, $\phi_\beta(X)$. Let ρ be the representation of $\mathfrak{g} \otimes \mathbb{C}$ on V. By Proposition 2.1, $\rho(\mathfrak{m})v_\beta = 0$ and by definition $\rho(v_i)v_\beta = 0$. Moreover, for $i \leq N'$, $\rho(H_i)v_\beta = \beta(H_i)v_\beta = 0$ since β is perpendicular to H_i by Lemma 5.1. To show that $\rho(\bar{v}_i)v_\beta = 0$ for $i \leq N'$ we need

LEMMA 8.5. *Let* v^+, v^- *and* z *be the standard basis of* $sL(2, \mathbb{R})$ *and let* V_o *be an irreducible* $sL(2, \mathbb{R})$*-module. Suppose* V_o *contains a non-zero vector which is annihilated by both* v^+ *and* z. *Then* V_o *is one-dimensional and* $sL(2, \mathbb{R})$ *acts trivially on it.*

Proof. This is a corollary of Lemma 5.3 of §5.

Now $v_i, \bar{v}_i$ and $\sqrt{-1}\,H_i$ span a subalgebra of $\mathfrak{g} \otimes \mathbb{C}$ isomorphic to $sL(2, \mathbb{R})$. Let us decompose V into a direct sum of irreducible $sL(2, \mathbb{R})$ modules

$$V = \sum V^{(k)}$$

and let $v_\beta = \sum v_\beta^{(k)}$ with $v_\beta^{(k)} \in V^{(k)}$: Since $\rho(v_i)v_\beta^{(k)} = \rho(H_i)v_\beta^{(k)} = 0$, we conclude from the lemma that either $v_\beta^{(k)} = 0$ or $v_\beta^{(k)}$ is annihilated by $\bar{v}_i$. This shows that v_β is annihilated by $\bar{v}_i$ and concludes the proof of Proposition 8.4. Q.E.D.

We recall that X_S is the orbit of G_S in X containing the identity coset. Let $K_S = G_S \cap K$. We can identify X_S with the homogeneous space, G_S/K_S. By (8.1) $\mathfrak{k} \cap \mathfrak{g}_S \subset \mathfrak{h}_S$ so we have

LEMMA 8.6. *The connected component of* K_S *is contained in* H_S.

COROLLARY. $H_S K_S$ *is a closed normal subgroup of* G_S *of the same dimension as* H_S *and the map*

$$G_S/H_S \to G_S/H_S K_S$$

is a finite covering map.

We will denote by Y_S the quotient space, $G_S/H_S K_S$. By Lemma 8.3, Y_S is a compact abelian group, i.e. a torus. Since $X_S = G_S/K_S$ we have a fiber mapping,

$$\tau : X_S \to Y_S . \tag{8.3}$$

We recall from §5 that $\mathfrak{a}$ contains a natural lattice, $\mathcal{L}$. As in (8.2) let $\mathfrak{a}_S$ be the subalgebra of $\mathfrak{a}$ spanned by $\{a, a \in \mathfrak{a}_S\}$ and let $\mathcal{L}_S = \mathfrak{a}_S \cap \mathcal{L}$.

LEMMA 8.7. $\mathcal{L}_S$ *is a co-compact lattice in* $\mathfrak{a}_S$ *and the quotient of* $\mathfrak{a}_S$ *by* $\mathcal{L}_S$ *is identical with* Y_S.

We will omit the proof.

Let $\mathcal{L}_S^*$ be the dual lattice to $\mathcal{L}_S$ in $\mathfrak{a}_S$. Given $\beta \in \mathcal{L}_S^*$ the function, $e^{i(\beta,a)}$, on $\mathfrak{a}$ is $\mathcal{L}$-periodic; so it defines a function, e_β, on Y_S. Conversely, every exponential function on Y_S is of this form.

We can now describe quite precisely the micro-local structure of the projector, π^S. To get a global result we will do what we did in §5 and kill off the contribution to π^S coming from the lattice points on ∂S. Let f be a smooth function on S which is homogeneous of degree zero and has support contained in the complement of a neighborhood of ∂S. If V is an irreducible subrepresentation of $L^2(X)$ with maximal weight $\beta \in \overline{S}$, we define π_f^S on V by setting, $\pi_f^S v_\beta = f(\beta) v_\beta$, v_β being the maximal weight vector, and $\pi_f^S = 0$ on the orthogonal complement of v_β in V.

This defines π_f^S on all irreducible subspaces of the type above. On all other irreducible subspaces of $L^2(X)$ we set $\pi_f^S \equiv 0$.

THEOREM 8.8. *The wave-front set of* π_f^S *is the set of all points of the form*

$$\{(x,\xi,y,\xi),\ \xi \in \text{support } f,\ x,y \in X_S,\ \tau(x)=\tau(y)\},$$

τ *being the mapping* (8.3).

We will denote by Σ_S^D the set

$$\{(x,\xi,y,\xi),\ \xi \in S,\ x,y \in X_S,\ \tau(x)=\tau(y)\}. \tag{8.4}$$

Theorem 8.8 says that for all f the wave-front set of π_f^S is contained in Σ_S^D. Note that Σ_S^D is usually much smaller than the set, Σ_S^Δ, of Proposition 8.1.

In the paragraph below we will use the following convenient notational convention. We will denote by $-T^*X$ the cotangent bundle of X, but with the reverse of its usual symplectic structure; i.e. with the symplectic structure given by the one-form, $\sum -\xi_i\,dx_i$.

LEMMA 8.9. Σ_S^D *is an isotropic submanifold of* $(T^*X)\times(-T^*X)$; *i.e. the product symplectic form restricts to be zero on* Σ_S^D.

Proof. LTR.

The following is an analogue of Theorem 5.7.

THEOREM 8.10. π_f^S *belongs to* $OP\mathcal{H}^0(X\times X, \Sigma_S^D)$.

For notation, see [1].

Let f be a smooth homogeneous function on S which vanishes near ∂S as above and, for each $\beta \in \mathfrak{L}_S^* \cap S$ set

(8.5) $$U_f^S e_\beta = f(\beta)\phi_\beta ,$$

e_β being the exponential function $e^{i(\beta,y)}$ and ϕ_β the conical function indexed by β. By (8.5) we can extend U_f^S to all of $L^2(Y_S)$. As in (6.5) let

(8.6) $$j : T^*X_S \to T^*X$$

be the symplectic imbedding given by the Riemannian metrices on X_S and X and let Λ_S be the isotropic submanifold of $(T^*Y_S) \times (-T^*X_S)$ consisting of the points

(8.7) $$(y, \eta, x, \xi), \quad x \in X_S, \quad y = \tau(x), \quad \xi = j(d\tau_x)^*\eta .$$

The following is an analogue of Proposition 7.4.

THEOREM 8.11. *U_f is in $OP\mathcal{H}^0(Y_S \times X, \Lambda_S)$.*

The proof of the Theorems 8.8, 8.10 and 8.11 will be given elsewhere. We will, however, give a few indications here about how the proofs go. Consider the system of equations

(8.8) $$D_{v_i} u = 0, \qquad i = 1, \cdots, N ,$$

on X the v_i's being as in Lemma 6.1. We recall from §3 that Σ_S is one component of the characteristic variety of this system. On this component the Levi form is a diagonal matrix with non-negative entries along the diagonal as before; however, *precisely* N′ *entries along the diagonal are zero,* namely the first N′ diagonal entries, since

$$1/\sqrt{-1}\,(\xi, [v_i, \overline{v}_i]) = (\xi, H_i) = 0$$

when $\xi \in S$ and $i \leq N'$, H_i being perpendicular to S by Lemma 5.1. To prove Theorem 8.10 (from which the other theorems follow by functorial nonsense) one has to resort to analogues of the Boutet-Sjostrand results described in §4 for systems with positive semi-definite Levi forms of *constant nullity*. We will discuss the micro-local theory of such systems in a forthcoming article.

We conclude with a couple of results on the asymptotic behavior of the conical functions, ϕ_β, $\beta \in S$ as β tends to infinity in Int S.

THEOREM 8.12. *Let* $\mathcal{W}$ *be a conic open subset of* S *whose closure in* $\mathfrak{a}_+ - 0$ *is disjoint from* ∂S, *and let* Q *be a zeroth order pseudodifferential operator on* X *with symbol* $\sigma(Q)$. *Then*

$$\int_X Q\phi_\beta\bar{\phi}_\beta dx = \int_{X_S} \sigma(Q)(x,\beta)\,dx + 0(|\beta|^{-1})$$

for all $\beta \in \mathcal{W}$.

COROLLARY. *As* β *tends to infinity in* $\mathcal{W}$, $|\phi_\beta|^2$ *tends to the delta function of* X_S.

BIBLIOGRAPHY

[1] L. Boutet de Monvel, "Hypoelliptic operators with double characteristics and related pseudodifferential operators," Comm. Pure Appl. Math. 27 (1974), 585-639.

[2] L. Boutet de Monvel and J. Sjostrand, "Sur la singularité des noyaux de Bergman et de Szego," Asterisque 134-35 (1976), 123-164.

[3] Y. Colin de Verdiere, "Quasi-modes sur les variétés riemanniennes," Inventiones Math. (to appear).

[4] V. Guillemin, "Symplectic spinors and partial differential equations," Colloque international de géométrie symplectique, Aix (June 1974) C.N.R.S.

[5] V. Guillemin and S. Sternberg, "On the spectra of commuting pseudodifferential operators: recent work of Kac-Spencer, Weinstein and others," Utah conference on analysis, (February 1977) (to appear).

[6] V. Guillemin and A. Weinstein, "Eigenvalues associated with a closed geodesic," Bulletin of the AMS, Vol. 82, no. 1 (Jan. 1976), 92-94.

[7] S. Helgason, *Differential geometry and symmetric spaces*, Academic Press, New York (1962).

[8] ———, *Analysis of Lie groups and homogeneous spaces*, Regional conference series in mathematics, no. 14, AMS, Providence (1972).

[9] L. Hormander, "Fourier Integral operators I," Acta Math. 127 (1971), 79-183.

[10] J. Humphreys, *Introduction to Lie algebras and representation theory*, Springer Verlag, New York (1972).

[11] N. Jacobson, *Lie algebras*, Interscience, New York (1962).

[12] G. Warner, *Harmonic analysis on semi-simple Lie groups*, Springer Verlag, New York (1972).

ON HOLONOMIC SYSTEMS WITH REGULAR SINGULARITIES

Masaki Kashiwara* and Takahiro Kawai*

A holonomic system, i.e., a left coherent $\mathcal{E}$-Module (or $\mathcal{D}$-Module)[(1)] whose characteristic variety is Lagrangian, shares the finiteness theorem with ordinary differential equations, namely, all the cohomology groups associated with its solution sheaf are finite dimensional ([4], [8]). Hence the study of a holonomic system will, in principle, give us almost complete information concerning the functions which satisfy the system, as in the one-dimensional case. When we try to work out such a program, concentrating our attention on holonomic systems "with regular singularities" will be a natural choice. The purpose of this report is to summarize the article of Kashiwara-Kawai [10], which establishes a solid basis for the theory and, at the same time, clarifies the role that holonomic systems with regular singularities play among general holonomic systems. (See Theorem 1 below for the precise statement.)

Before beginning the discussions, we mention that our work is closely related to the work of Nilsson [17], Leray [13], and Deligne [2]. Especially our argument makes essential use of the results of Deligne [2]. A very

---

*Supported in part by NSF grant MCS77-18723

(1) $\mathcal{E}$ (resp. $\mathcal{D}$) denotes the sheaf of micro-differential (resp. linear differential) operators of finite order.

Seminar on Micro-Local Analysis
0-691-08228-6/79/00 0113-09 $00.50/1 (cloth)
0-691-08232-4/79/00 0113-09 $00.50/1 (paperback)

interesting paper of Ramis [18] is also closely related to a part of our results.(2)

In this report we use the same notions and notations as in [10]. See also [11], [19].

We first recall some notions needed to define systems with regular singularities ([11]).

We denote by $\mathcal{E}_X(m)$ the sheaf of micro-differential operators of order at most m. Let V be a homogeneous involutory subvariety (possibly with singularities) of T^*X and I_V the sheaf of holomorphic functions on T^*X which vanish on V. Then we denote by $\mathcal{J}_V$ the sheaf $\{P \in \mathcal{E}_X(1); \sigma_1(P) \in I_V\}$ and by $\mathcal{E}_V$ the sub-Algebra of $\mathcal{E}_X$ generated by $\mathcal{J}_V$. Using the sheaf $\mathcal{E}_V$, we define the notion of $\mathcal{E}_X$-Module with regular singularities along V as follows:

DEFINITION 1. Let $\mathcal{M}$ be a coherent $\mathcal{E}_X$-Module defined on $\Omega \subset T^*X$. We say that $\mathcal{M}$ has regular singularities along V if one of the following three equivalent conditions is satisfied:

(i) For any point p in Ω, there exists a neighborhood U of p and an $\mathcal{E}_V$-sub-Module $\mathcal{M}_0$ of $\mathcal{M}$ defined on U which is coherent over $\mathcal{E}(0)$ and which generates $\mathcal{M}$ as $\mathcal{E}_X$-Module, i.e., $\mathcal{M} = \mathcal{E}_X \mathcal{M}_0$.

(ii) For any coherent $\mathcal{E}(0)$-sub-Module $\mathcal{L}$ of $\mathcal{M}$ defined on an open set of Ω, $\mathcal{E}_V \mathcal{L}$ is coherent over $\mathcal{E}(0)$.

(iii) Let $\mathcal{N}$ be an $\mathcal{E}_V$-sub-Module of $\mathcal{M}$ that is defined on an open set of Ω and that is locally of finite type over $\mathcal{E}_V$, i.e., locally generated by finitely many sections of $\mathcal{E}_V$. Then $\mathcal{N}$ is coherent over $\mathcal{E}(0)$.

DEFINITION 2. For an $\mathcal{E}_X$-Module $\mathcal{M}$ and an involutory variety V which contains $\mathrm{Supp}\,\mathcal{M}$, $IR(\mathcal{M}, V)$ denotes the set $\{p; \mathcal{M}$ does not have regular singularities along V on any neighborhood of $p\}$.

---

(2) Even though the definition of "regular singularities" of Ramis [18] is different from ours, an analytic characterization ([10], Chap. VI) of holonomic $\mathcal{D}$-Modules with regular singularities shows that they are actually the same.

DEFINITION 3. A holonomic $\mathcal{E}_X$-Module $\mathfrak{M}$ is said to be with R.S. if and only if $\mathrm{Supp}\,\mathfrak{M} \cap \mathrm{IR}(\mathfrak{M}, \mathrm{Supp}\,\mathfrak{M})$ is nowhere dense in $\mathrm{Supp}\,\mathfrak{M}$.

REMARK 1. Note that we have defined the notion of a holonomic system with R.S. by a property of the system at the generic points of its characteristic variety. However, we can eventually prove ([10], Chap. V) that $\mathfrak{M}$ has regular singularities along V for any involutory set V containing $\mathrm{Supp}\,\mathfrak{M}$ if $\mathfrak{M}$ is a holonomic system with R.S.

DEFINITION 4. For a holonomic $\mathcal{E}_X$-Module $\mathfrak{M}$ we define a subsheaf $\mathfrak{M}_{reg}$ of $\mathfrak{M}^\infty \equiv \mathcal{E}^\infty \otimes \mathfrak{M}$ by defining the presheaf $\{\mathfrak{M}_{reg}(U)\}$ by $\mathfrak{M}_{reg}(U) = \{s \in \mathfrak{M}^\infty(U);$ for any point p in U there exists an Ideal $\mathcal{J} \subset \mathcal{E}$ defined near p such that $\mathcal{J}s = 0$ and that $\mathcal{E}/\mathcal{J}$ is with R.S.$\}$.

REMARK 2. Making use of the detailed analysis on the structure of $\mathfrak{M}$ at the generic points of $\mathrm{Supp}\,\mathfrak{M}$, we can prove ([10], Chap. I, §3) that $\mathfrak{M}_{reg}$ is a holonomic $\mathcal{E}_X$-Module with R.S.

Now, one of the most important results of [10] is the following:

THEOREM 1. *For any holonomic $\mathcal{E}$-Module $\mathfrak{M}$, $\mathcal{E}^\infty \underset{\mathcal{E}}{\otimes} \mathfrak{M}_{reg} = \mathcal{E}^\infty \underset{\mathcal{E}}{\otimes} \mathfrak{M}$ holds.*

In view of Remark 2, this theorem asserts that any holonomic $\mathcal{E}$-Module can be transformed into a holonomic $\mathcal{E}$-Module with R.S. by micro-differential operators of infinite order. As far as we know, such a clear result has not been known even for ordinary differential equations, even though several transformations are employed in analyzing equations with irregular singularities (by Birkhoff, Hukuhara, Turrittin, $\cdots$).

The proof of this theorem is achieved by constructing sufficiently many multi-valued analytic solutions of the system $\mathfrak{M}$ so that we can imbed $\mathfrak{M}$ into a $\mathcal{D}$-Module. Here we essentially use two results of Deligne [2]: The first is the result to the effect that for any multi-valued analytic function ϕ with finite determination we can find a Nilsson class function ψ with the same monodromy structure as ϕ. As a matter of fact, what we need is a mcre sophisticated version of this result which makes use of linear differential operators of infinite order. It follows from "Reconstruction Theorem," i.e., a theorem which establishes the exact correspondence between the category of holonomic $\mathcal{D}_X$-Modules and the category of constructible sheaves on X. See [10], Chap. I, §4 and Chap. II, §2 for details. The second is the result to the effect that a Nilsson class function satisfies a holonomic system of linear differential equations. (See also Kashiwara [6].) The complete proof of Theorem 1 is too long and complicated to reproduce here and we refer the reader to [10] for it. A prototype of the argument can be found in Kashiwara-Kawai [9]. See also Kashiwara-Kawai [7], [8] and Bony-Schapira [1]. A recent result of Kashiwara-Schapira [12] on micro-hyperbolic systems is effectively used in the proof.

In the course of the proof of Theorem 1 we find the following Theorem 2, which is also very interesting and important.

THEOREM 2. *Let* $\mathfrak{M}$ *be a holonomic* $\mathcal{E}_X$*-Module with* R.S. *Assume that* Supp $\mathfrak{M}$ *is in a generic position near* p, *namely,* Supp $\mathfrak{M} \cap \pi^{-1}\pi(p) = \mathbb{C}^\times p$(3) *in a neighborhood of* p. *Then* $\mathfrak{M}_p$ *is a* $\mathcal{D}_{X,\pi(p)}$*-module*(4) *of finite type and satisfies the following*:

$$\mathcal{E}_{X,q} \underset{\pi^{-1}\mathcal{D}_{X,\pi(p)}}{\otimes} \mathfrak{M}_p = \begin{cases} \mathfrak{M}_p & \text{if} \quad q = p \\ 0 & \text{if} \quad q \in \pi^{-1}\pi(p) - T^*_X X - \mathbb{C}^\times p \,. \end{cases}$$

REMARK 3. We conjecture that this result holds for any holonomic $\mathcal{E}_X$-Module with R.S.

---

(3) $\mathbb{C}^\times$ denotes the multiplicative group of non-zero complex numbers and $\mathbb{C}^\times p$ means $\{(x, c\xi);\ c \in \mathbb{C}^\times\}$ with $p = (x, \xi)$.

(4) π denotes the canonical projection from T^*X to X.

Using these results we can establish the fact that the family of holonomic systems with R.S. is closed under integration procedure and restriction procedure. More precisely we have the following results. (See [10], Chap. V. See also [19], Chap. III, §3.5 and §4.2 and [5], §4 for related topics.)

THEOREM 3. *Let* $\phi : Y \to X$ *be a holomorphic map and* $\mathfrak{M}$ *a holonomic* $\mathcal{E}_X$*-Module with* R.S. *defined on an open set* U *in* $T^*X - T^*_X X$. *Let* W *be an open set in* $T^*Y - T^*_Y Y$ *such that* $\varpi^{-1}(\mathrm{Supp}\,\mathfrak{M}) \cap \rho^{-1}(W) \to W$ *is finite. Then* $\phi^*\mathfrak{M} \equiv \rho_*(\mathcal{E}_{Y\to X} \underset{\varpi^{-1}\mathcal{E}_X}{\otimes} \varpi^{-1}\mathfrak{M})$ *is a holonomic* $\mathcal{E}_Y$*-Module with* R.S. *defined on* W. *Here and in the sequel* ϖ *(resp.* ρ*) denotes the canonical projection from* $Y \underset{X}{\times} T^*X$ *to* T^*X *(resp.* T^*Y*).*

THEOREM 4. *Let* $\phi : Y \to X$ *be a holomorphic map,* U *an open set in* $T^*X - T^*_X X$ *and* W *an open set in* $T^*Y - T^*_Y Y$. *Let* $\mathfrak{N}$ *be a coherent right* $\mathcal{E}_Y$*-Module defined on* W. *Assume that* $\mathfrak{N}$ *is a holonomic system with* R.S. *Assume furthermore that* $\rho^{-1}\,\mathrm{Supp}\,\mathfrak{N} \cap \varpi^{-1}(U) \to U$ *is finite. Then* $\phi_*\mathfrak{N} \equiv \varpi_*(\rho^{-1}\mathfrak{N} \underset{\rho^{-1}\mathcal{E}_Y}{\otimes} \mathcal{E}_{Y\to X})$ *is a right holonomic* $\mathcal{E}_X$*-Module with* R.S. *defined on* U.

If we assume in addition that $\mathfrak{M}$ is a $\mathcal{D}_X$-Module, we can generalize these results as follows:

THEOREM 5. *Let* $\phi : Y \to X$ *be a holomorphic map and* $\mathfrak{M}$ *a holonomic* $\mathcal{D}_X$*-Module with* R.S. *Then* $\phi^*\mathfrak{M} = \mathcal{O}_Y \underset{f^{-1}\mathcal{O}_X}{\otimes} f^{-1}\mathfrak{M}$ *is a holonomic* $\mathcal{D}_Y$*-Module with* R.S.

THEOREM 6. *Let* $\phi : Y \to X$ *be a projective map and* $\mathfrak{M}$ *a holonomic* $\mathcal{D}_Y$*-Module with* R.S. *Then, for any* k, $\int^k \mathfrak{M} \equiv R^k\phi_*(\mathcal{D}_{X\leftarrow Y} \overset{L}{\underset{\mathcal{D}_Y}{\otimes}} \mathfrak{M})$ *is a holonomic* $\mathcal{D}_X$*-Module with* R.S.

REMARK 4. Since Theorem 6 gives information on the characteristic variety of $\int^k \mathcal{M}$, it will be useful in manipulating Nilsson class functions.

Next we shall discuss how we can 'analytically' characterize holonomic systems with regular singularities. The basic properties of holonomic systems with R.S. stated so far are effectively used for this purpose ([10], Chap. V and Chap. VI).

Let us first recall the most important characteristic property of the ordinary differential equations with regular singularities, namely, the validity of the comparison theorem of the following type.

THEOREM 7 (Malgrange [15]). *Let* $\mathcal{M}$ *be* $\mathcal{D}_{\mathbf{C}}/\mathcal{D}_{\mathbf{C}}P$ $(P \in \mathcal{D}_{\mathbf{C},0})$. *Assume* $SS(\mathcal{M}) = T^*_{\{0\}}\mathbf{C} \cup T^*_{\mathbf{C}}\mathbf{C}$. *Then* $\mathcal{M}$ *has regular singularities at* 0 *if and only if*

(1)
$$\mathcal{E}xt^j_{\mathcal{D}_{\mathbf{C}}}(\mathcal{M}, \mathcal{O}_{\mathbf{C}})_0 = \mathcal{E}xt^j_{\mathcal{D}_{\mathbf{C}}}(\mathcal{M}, \hat{\mathcal{O}}_{\mathbf{C},0})$$

holds for $j = 0, 1$. *Here* $\hat{\mathcal{O}}_{\mathbf{C},0} = \varprojlim_k \mathcal{O}_{\mathbf{C},0}/\mathfrak{m}^k$, *where* $\mathfrak{m}$ *is the maximal ideal of* $\mathcal{O}_{\mathbf{C},0}$.

It is noteworthy that such a basic result had not been obtained apparently before Malgrange [15]. Probably this is due to the fact that the characterization of regular singularities requires not only the study of the 0-th cohomology group but also that of the first cohomology group,(5) while specialists in the theory of ordinary differential equations had rarely considered higher order cohomology groups before Deligne [2].

---

(5) In order to illustrate this, we consider the equation $\mathcal{M} = \mathcal{D}_{\mathbf{C}}/\mathcal{D}_{\mathbf{C}}P$ with $P = x^2 D_x - a (a \neq 0)$. Clearly $\mathcal{M}$ is *not* with regular singularity at the origin. However, $\mathcal{E}xt^0_{\mathcal{D}_{\mathbf{C}}}(\mathcal{M}, \mathcal{O}_{\mathbf{C}})_0 = \mathcal{E}xt^0_{\mathcal{D}_{\mathbf{C}}}(\mathcal{M}, \hat{\mathcal{O}}_{\mathbf{C},0})$ holds—actually, both hand sides are zero! See also Gérard-Sibuya [3] and Majima [14] for very interesting related results on a class of Pfaff systems.

The holonomic system with R.S. in our sense shares such comparison theorems with ordinary differential equations with regular singularities as follows:

THEOREM 8. *Let* $\mathcal{M}$ *and* $\mathcal{N}$ *be holonomic* $\mathcal{E}$*-Modules with* R.S. *Then*

$$\mathcal{E}xt^j(\mathcal{M}, \mathcal{N}) \simeq \mathcal{E}xt^j(\mathcal{M}, \mathcal{N}^\infty)$$

holds for any j.

THEOREM 9. *Let* $\mathcal{M}$ *be a holonomic* $\mathcal{D}_X$*-Module with* R.S. *Then*

$$\mathcal{D}_X^\infty \underset{\mathcal{D}_X}{\otimes} \mathcal{H}^j_{[Y]}(\mathcal{M}) \simeq \mathcal{H}^j_Y(\mathcal{D}_X^\infty \underset{\mathcal{D}_X}{\otimes} \mathcal{M})$$

for any j *and any analytic subset* Y *of* X. *Here* $\mathcal{H}^j_{[Y]}(\mathcal{M})$ *denotes the* j-th *algebraic relative cohomology group of* $\mathcal{M}$ *supported by* Y, *namely,* $\varinjlim_m \mathcal{E}xt^j_{\mathcal{O}_X}(\mathcal{O}_X/\mathcal{I}_Y^m, \mathcal{M})$, *where* $\mathcal{I}_Y$ *is the defining Ideal of* Y.

REMARK 5. A special case of Theorem 9 where $\mathcal{M} = \mathcal{O}_X$ was proved by Mebkhout [16]. Ramis [18] defines the notion of a fuchsian holonomic $\mathcal{D}$-Module by using this property as its characteristic property.

As a corollary of Theorem 8 we obtain the following Theorem 10, which justifies our usage of the terminology "with R.S."

THEOREM 10. *Let* $\mathcal{M}$ *be a holonomic* $\mathcal{D}_X$*-Module with* R.S. *Then*

$$\mathcal{E}xt^j_{\mathcal{D}_X}(\mathcal{M}, \mathcal{O}_X)_x \simeq \mathcal{E}xt^j_{\mathcal{D}_X}(\mathcal{M}, \hat{\mathcal{O}}_{X,x}) \tag{2}$$

holds for any j *and any* x *in the domain of definition of* $\mathcal{M}$. *Here* $\hat{\mathcal{O}}_{X,x} = \varprojlim_k \mathcal{O}_{X,x}/\mathfrak{m}^k$, *where* $\mathfrak{m}$ *is the maximal ideal of* $\mathcal{O}_{X,x}$.

Furthermore we can prove that the validity of (2) is actually a characteristic property of holonomic $\mathfrak{D}_X$-Module with R.S., namely, we have the following Theorem 11 as a generalization of Theorem 7.

THEOREM 11. *Let* $\mathfrak{M}$ *be a holonomic* $\mathfrak{D}_X$*-Module. Assume that* (2) *holds. Then* $\mathfrak{M}$ *is with* R.S.

REFERENCES

[1] Bony, J. M. and Schapira, P. Propagation des singularités analytiques pour les solutions des équations aux dérivées partielles. Ann. Inst. Fourier *26*, 81-140 (1976).

[2] Deligne, P. Equations Differentielles à Points Singuliers Réguliers. Lecture Notes in Math. No. 163, Berlin-Heidelberg-New York, Springer, 1970.

[3] Gérard, R. et Sibuya, Y. Étude de certains systèmes de Pfaff au voisinage d'une singularité. C. R. Acad. Sc. Paris *284*, 57-60 (1977).

[4] Kashiwara, M. On the maximally overdetermined system of linear differential equations, I. Publ. RIMS, Kyoto Univ. *10*, 563-579 (1975).

[5] ———. B-functions and holonomic systems. Inventiones math. *38*, 33-53 (1976).

[6] ———. On holonomic systems of linear differential equations II. To appear in Inventiones math.

[7] Kashiwara, M. and Kawai, T. Micro-hyperbolic pseudo-differential operators I. J. Math. Soc. Japan *27*, 359-404 (1975).

[8] ———. Finiteness theorem for holonomic systems of micro-differential equations. Proc. Japan Acad. *52*, 341-343 (1976).

[9] ———. Holonomic character and local monodromy structure of Feynman integrals. Commun. math. Phys. *54*, 121-134 (1977).

[10] ———. On holonomic systems of micro-differential equations III. Systems with regular singularities. To appear.

[11] Kashiwara, M. and Oshima, T. Systems of differential equations with regular singularities and their boundary value problems. Ann. of Math. *106*, 145-200 (1977).

[12] Kashiwara, M. and Schapira, P. Micro-hyperbolic systems. To appear in Acta Math.

[13] Leray, J. Un complément au théorème de N. Nilsson sur les integrales de formes differentielles à support singulier algebrique. Bull. Soc. Math. France *95*, 313-374 (1967).

[14] Majima, H. Remarques sur la théorie de development asymptotique en plusieurs variables I. Proc. Japan Acad. *54*, Ser. A. 67-72 (1978).

[15] Malgrange, B. Sur les points singuliers des équations différentielles. Enseignement Math. *20*, 147-176 (1974).

[16] Mebkhout, Z. Local cohomology of analytic spaces. Publ. RIMS, Kyoto Univ. *12*, Suppl. 247-256 (1977).

[17] Nilsson, N. Some growth and ramification properties of certain integrals on algebraic manifolds. Ark. Mat. *5*, 527-540 (1963-65).

[18] Ramis, J. P. Variations sur le theme "GAGA." To appear.

[19] Sato, M., Kawai, T. and Kashiwara, M. Microfunctions and pseudo-differential equations. Lecture Notes in Math. No. 287, Berlin-Heidelberg-New York, Springer, 1973, pp. 265-529.

MICRO-LOCAL ANALYSIS OF FEYNMAN AMPLITUDES
(Seminar Report on Linear PDE given in the fall of 1977)

Masaki Kashiwara* and Takahiro Kawai*

The purpose of this report is two-fold:

On the one hand, we show that Feynman integrals are an interesting object for mathematicians to study, especially from the viewpoint of micro-local analysis. On the other hand, we show that the micro-local analysis of Feynman integrals yields several physically interesting results on the analytic structure of Feynman integrals. Even though some results of this report can be extended to the S-matrix itself, we do not discuss it here. (See Kawai-Stapp [16], [17] for this topic.)

Although we do not give any explanations why the study of analyticity properties of the S-matrix and Feynman integrals are physically interesting and important, we just quote one sentence from the celebrated and pioneering book of Chew [3], where he claims "Analyticity as a fundamental principle in physics":

> "During the past ten years, nevertheless, a feeling has been growing among many theoretical physicists that the description of natural phenomena on subatomic level may be facilitated if analyticity is employed as a primary rather than a derived concept." (Chew [3], p. 2, line 33-line 36.)

*Supported in part by National Science Foundation grant MCS77-18723.

Seminar on Micro-Local Analysis
0-691-08228-6/79/00 0123-15 $00.75/1 (cloth)
0-691-08232-4/79/00 0123-15 $00.75/1 (paperback)

For the details of the results discussed here, we refer the reader to Kashiwara-Kawai [9], [10], [11]; Kashiwara-Kawai-Oshima [13]; Kashiwara-Kawai-Stapp [14] and Sato-Miwa-Jimbo-Oshima [27]. See Nakanishi [20], Speer [28] and Kawai-Stapp [17] for notations related to Feynman integrals and diagrams and see Sato-Kawai-Kashiwara [26] for the basic notions concerning microfunctions and holonomic (= maximally overdetermined) systems of micro-differential (= pseudo-differential) equations. See also Nakanishi [20], Speer [28] and references cited there for the physical importance of Feynman integrals.

First of all, we recall the definition of Feynman diagrams and Feynman integrals.

DEFINITION 1. Feynman diagram D consists of finitely many points (called "vertices") $\{V_j\}_{j=1,\cdots,n'}$, finitely many one-dimensional segments (called "internal lines") $\{L_\ell\}_{\ell=1,\cdots,N}$, and finitely many half lines (called "external lines") $\{L_r^e\}_{r=1,\cdots,n}$. Each of the end points W_ℓ^+ and W_ℓ^- of L_ℓ and the end point of L_r^e must coincide with some vertex V_j. We suppose $W_\ell^+ \neq W_\ell^-$ for all ℓ.(*) A four vector $p_r = (p_{r,0}, p_{r,1}, p_{r,2}, p_{r,3})$ is associated with each external line L_r^e and a strictly positive(*) constant m_ℓ(**) is associated with each internal line L_ℓ. We suppose that each internal line and each external line are oriented.(*) The orientation is indicated by the arrow $\longrightarrow$.

Example of a Feynman diagram D:

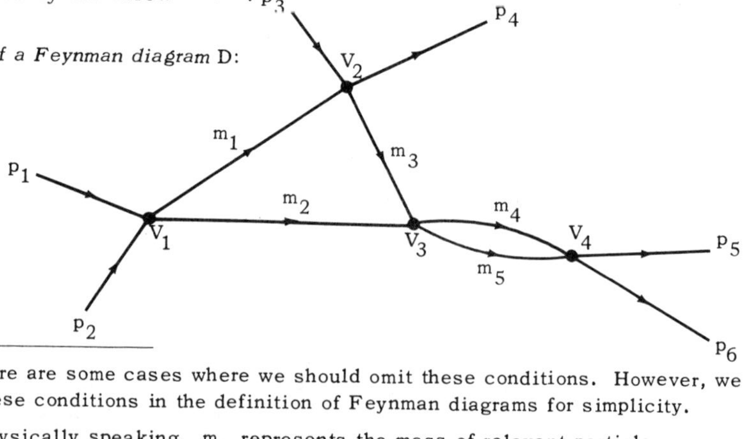

(*)There are some cases where we should omit these conditions. However, we include these conditions in the definition of Feynman diagrams for simplicity.

(**)Physically speaking, m_ℓ represents the mass of relevant particles.

REMARK. A Feynman diagram represents diagrammatically the interaction of elementary particles. See, e.g., Nakanishi [20] for more detailed physical explanations.

DEFINITION 2. (Incidence number). If internal line L_ℓ starts from V_j (resp. ends at V_j), the incidence number $[j:\ell]$ is defined to be -1 (resp. $+1$). In other cases, $[j:\ell]$ is defined to be zero. The incidence number $[j:r]$ is defined in the same way.

DEFINITION 3. A vertex V_j of D for which $[j:r]=0$ holds for all r is called internal vertex. Other vertices are called external.

REMARK. For simplicity all diagrams considered below are supposed to be connected.

DEFINITION 4. (Feynman rule)$^{(*)}$. Feynman integral $F_D(p)$ associated with D is (formally) defined by

$$(1)\ F_D(p) = F_D(p_1,\cdots,p_n) \equiv \int \frac{\prod_{j=1}^{n'} \delta^4\left(\sum_{r=1}^{n} [j:r]p_r + \sum_{\ell=1}^{N} [j:\ell]k_\ell\right)}{\prod_{\ell=1}^{N} (k_\ell^2 - m_\ell^2 + \sqrt{-1}\,0)} \prod_{\ell=1}^{N} d^4k_\ell\,,$$

where $k_\ell^2 = k_{\ell,0}^2 - \sum_{\nu=1}^{3} k_{\ell,\nu}{}^2$.$^{(**)}$

Since $F_D(p)$ "defined" by (1) is, in general, a divergent integral and not well defined as it stands, the so-called renormalization procedure is needed to make it well defined. The recipe for renormalization given by Bogoliubov-Parasiuk-Hepp is not convenient for our purposes; we use here

(*)For simplicity we suppose all relevant particles are spinless.

(**)In this report, we always use this Minkowsky metric to define k^2 for a four-vector k.

the recipe given by Speer [28] (which is equivalent to that given by Bogoliubov-Parasiuk-Hepp aside from finite renormalization). The recipe of Speer is as follows:(*)

First we consider the generalized Feynman integral $F_D(p;\lambda)$ defined as follows:

$$(2) \qquad F_D(p;\lambda) = F_D(p_1,\cdots,p_n;\lambda_1,\cdots,\lambda_N)$$

$$\equiv \int \frac{\prod_{j=1}^{n'} \delta^4\left(\sum_{r=1}^{n} [j:r]p_r + \sum_{\ell=1}^{N} [j:\ell]k_\ell\right)}{\prod_{\ell=1}^{N} (k_\ell^2 - m_\ell^2 + \sqrt{-1}\,0)^{\lambda_\ell}} \prod_{\ell=1}^{N} d^4 k_\ell .$$

This integral is convergent for $\operatorname{Re}\lambda_\ell \gg 1$ and meromorphically extended with respect to $\lambda (\in \mathbf{C}^N)$. Next we choose positive numbers $R_\ell\ (\ell = 1,\cdots,N)$ so that

$$(3) \qquad 0 < R_1 < R_2 < \cdots < R_N \ll 1$$

and

$$(4) \qquad R_\ell > \sum_{k=1}^{\ell-1} R_k$$

hold. We define $C_\ell = \{\lambda \in \mathbf{C};\ |\lambda - 1| = R_\ell\}$. Then the renormalized integral $F_D(p) \equiv \mathfrak{W}_N(F_D(p;\lambda))$ is given by

$$(5) \qquad \frac{1}{N!} \sum_{\sigma \in \mathfrak{S}_N} \oint_{C_{\sigma(1)}} \cdots \oint_{C_{\sigma(N)}} \frac{F_D(p;\lambda)}{(\lambda_1 - 1)\cdots(\lambda_N - 1)}\, d\lambda_1 \cdots d\lambda_N ,$$

where σ is a permutation of $(1,\cdots,N)$.

(*) Even though Speer's setting is more general, we have chosen the most convenient recipe for our purpose.

For our later discussions it is more convenient if we deal with integral (2) by compactifying the domain of integration, namely, we rewrite (2) as follows:

$$(6)\quad \int_{(P(R^4))^N} \frac{\prod_{j=1}^{n'} \delta^4\left(\sum_{r=1}^{n}[j:r] \prod_{\{\ell';[j:\ell']\neq 0\}} c_{\ell'} p_r + \sum_{\ell=1}^{N}[j:\ell] \frac{\prod_{\{\ell';[j:\ell']\neq 0\}} c_{\ell'}}{c_\ell} K_\ell\right)}{\prod_{\ell=1}^{N} (K_\ell^2 - c_\ell^2 m_\ell^2 + \sqrt{-1}\,0)^{\lambda_\ell}}$$

$$\times \prod_{\ell=1}^{N} |c_\ell|^{\mu_\ell} \omega(K_1, c_1)\cdots\omega(K_N, c_N).\ ^{(*)}$$

Here $\mu_\ell = 2\lambda_\ell + 4\#\{j;\ [j:\ell]\neq 0\} - 5 = 2\lambda_\ell + 3$, $(K_\ell, c_\ell)_{1\leq \ell\leq N}$ is the homogeneous coordinate on $P(R^4)$, namely, $K_\ell/c_\ell = k_\ell$, and $\omega(K_\ell, c_\ell)$

$$\equiv c_\ell^5\, d\left(\frac{K_{\ell,0}}{c_\ell}\right) d\left(\frac{K_{\ell,1}}{c_\ell}\right) d\left(\frac{K_{\ell,2}}{c_\ell}\right) d\left(\frac{K_{\ell,3}}{c_\ell}\right)$$

$$= \sum_{\nu=0}^{3} (-1)^\nu K_{\ell,\nu} dK_{\ell,0}\wedge\cdots\wedge dK_{\ell,\nu-1}\wedge dK_{\ell,\nu+1}\wedge\cdots\wedge dK_{\ell,3}\wedge dc_\ell + c_\ell dK_{\ell,0}\wedge\cdots\wedge dK_{\ell,3}$$

is the volume element on $P(R^4)$. (Sato-Miwa-Jimbo-Oshima [27], Jimbo [8].)

Our starting point is to know how the singularity spectrum of $F_D(p;\lambda)$ is described. As a matter of fact, it is described by the (positive $-\alpha$) Landau-Nakanishi variety defined below. (Landau [18], Nakanishi [19] and unpublished article of Bjorken.)

(*) The factor $\prod_{\{\ell';[j:\ell']\neq 0\}} c_{\ell'}$ is multiplied in the argument of δ^4 to make it well defined. Note that $\prod_{j=1}^{n'} \prod_{\{\ell';[j:\ell']\neq 0\}} c_{\ell'} = \prod_{\ell=1}^{N} c_\ell^2$ holds, because each c_ℓ appears exactly twice in the left-hand side.

DEFINITION 5. An 8n-real vector $(p, u) \equiv (p_1, \cdots, p_n, u_1, \cdots, u_n)$ is said to satisfy the positive $-\alpha$ Landau-Nakanishi equations if it satisfies the following equations for some $\alpha_\ell \in \overline{R^+} = \{\alpha \in R; \alpha \geq 0\}$, $w_j \in R^4$ and $k_\ell \in R^4$.

$$(7) \quad \begin{cases} u_r = \sum_{j=1}^{n'} [j:r] w_j & (r = 1, \cdots, n) \quad (7.a) \\ \sum_{r=1}^{n} [j:r] p_r + \sum_{\ell=1}^{N} [j:\ell] k_\ell = 0 & (j = 1, \cdots, n') \quad (7.b) \\ \sum_{j=1}^{n'} [j:\ell] w_j = \alpha_\ell k_\ell & (\ell = 1, \cdots, N) \quad (7.c) \\ \alpha_\ell (k_\ell^2 - m_\ell^2) = 0 & (\ell = 1, \cdots, N) \quad (7.d) \end{cases}$$

The positive$-\alpha$ Landau-Nakanishi variety $\mathcal{L}(D^+)$ is defined to be the set of points $(p, \sqrt{-1}\, u\infty) \in \sqrt{-1}\, S^* R^{4n}$, where (p, u) is a solution of (7).

REMARK 1. Since $F_D(p)$ (resp. $F_D(p; \lambda)$) has the form $\delta^4\left(\sum_{j,r} [j:r] p_r\right) f_D(p)$ (resp. $\delta^4\left(\sum_{j,r} [j:r] p_r\right) f_D(p; \lambda)$), we often discuss the analyticity property of $f_D(p)$ on $M = \left\{\sum_{j,r} [j:r] p_r = 0\right\} \subset R^{4n}$. By the standard convention that (p, u) and (p', u') represent the same point in S^*M if and only if $p = p'$ and $u - u' = a \in R^4$, we may regard $\mathcal{L}(D^+)$ as a subset in $\sqrt{-1}\, S^*M$. This convention will be often employed without explicit mentioning, if there is no fear of confusion. The function $f_D(p)$ (resp. $f_D(p; \lambda)$) is called a Feynman amplitude (resp. a generalized Feynman amplitude).

REMARK 2. We denote by $\mathcal{L}(D)$ the variety defined by the equation (7) without the additional assumption that $\alpha_\ell \geq 0$. The celebrated result of Landau and Nakanishi claims that the singularity of Feynman amplitude $f_D(p)$ is described by positive$-\alpha$ Landau-Nakanishi equations. Their result can, actually, be formulated micro-locally as follows:

THEOREM 1. $S.S.F_D(p) \subset \mathcal{L}(D^+)$.

For the rigorous proof, see Chandler [1] and Sato-Miwa-Jimbo-Oshima [27]. See also Chandler-Stapp [2], Iagolnitzer-Stapp [5], Pham [22], Sato [25], Iagolnitzer [5], [6], Kawai-Stapp [16], [17] and references cited there for related topics.

Concerning the analytic structure of $F_D(p)$, Regge [23] and Sato [25] made the following important and intriguing conjecture.

CONJECTURE. $F_D(p)$ satisfies a holonomic system of micro-differential equations whose characteristic variety is contained in $\mathcal{L}(D)^{\mathbf{C}}$, the complexification of $\mathcal{L}(D)$.

This conjecture has been proved with a slight modification concerning its characteristic variety by Kashiwara-Kawai [9]. The study on

$\prod_{\ell=1}^{N} (f_\ell + \sqrt{-1}\, 0)^{\lambda_\ell}$ done by Kashiwara-Kawai [11] is essential for the proof.

In order to state the result we recall the definition of extended Landau variety $\tilde{\mathcal{L}}(D)$. This variety appears naturally not only in mathematical context but also in physical context. (See Kashiwara-Kawai-Stapp [14].)

DEFINITION 6. $\tilde{\mathcal{L}}(D) \equiv \{(p, u) \in T^*\mathbf{C}^n$; there exists a sequence of scalars $c_\ell^{(m)}$ and $\alpha_\ell^{(m)}$ $(\ell = 1, \cdots, N)$ and four-vectors $p_r^{(m)}, u_r^{(m)}$ $(r = 1, \cdots, n)$, $k_\ell^{(m)}$ $(\ell = 1, \cdots, N)$ and $w_j^{(m)}$ $(j = 1, \cdots, n')$ which satisfies the following relation (8)$\}$.

$$
(8)\quad \begin{cases}
p_r^{(m)} \to p_r & (r=1,\cdots,n) \quad (8.a)\\
u_r^{(m)} \to u_r & (r=1,\cdots,n) \quad (8.b)\\
u_r^{(m)} = \sum_{j=1}^{n'} [j:r] w_j^{(m)} & (r=1,\cdots,n) \quad (8.c)\\
\sum_{r=1}^{n} [j:r] p_r^{(m)} + \sum_{\ell=1}^{N} [j:\ell] k_\ell^{(m)} = 0 & (j=1,\cdots,n') \quad (8.d)\\
\dfrac{\sum_{j=1}^{n'} [j:\ell] w_j^{(m)} + \alpha_\ell^{(m)} k_\ell^{(m)}}{c_\ell^{(m)}} \to 0 & (\ell=1,\cdots,N) \quad (8.e)\\
\alpha_\ell^{(m)} (k_\ell^{(m)2} - m_\ell^2) \to 0 & (\ell=1,\cdots,N) \quad (8.f)\\
c_\ell^{(m)} \text{ is bounded} & (\ell=1,\cdots,N) \quad (8.g)\\
c_\ell^{(m)} k_\ell^{(m)} \text{ is bounded} & (\ell=1,\cdots,N) \quad (8.h)\\
(c_\ell^{(m)}, c_\ell^{(m)} k_\ell^{(m)}) \not\to 0 & (\ell=1,\cdots,N) \quad (8.i)
\end{cases}
$$

Using $\tilde{\mathfrak{L}}(D)$, the result of [9] is stated as follows.

THEOREM 2. *Renormalized Feynman integral* $F_D(p)$ *satisfies a holonomic system* $\mathfrak{M}_D$ *of linear differential equations whose characteristic variety is contained in the extended Landau variety* $\tilde{\mathfrak{L}}(D)$.

Furthermore a recent result of Kashiwara-Kawai [12] shows that the system $\mathfrak{M}_D$ is actually a system with regular singularities. This property will turn out to be important when one tries to relate "holonomic system approach" to "monodromy structure approach." (See Regge [24] and references cited there for related topics.)

Even though a detailed study on the structure of $\mathfrak{M}_D$ has not yet been fully done in general, several interesting pieces of information on $F_D(p)$ can be drawn from the analysis of $\mathfrak{M}_D$ if we assume some moderate conditions on D. (Kashiwara-Kawai [10], Kashiwara-Kawai-Oshima [13].) As an example, we determine the singularity structure of $f_D(p)$ explicitly at some physically important points.

For the sake of simplicity, we consider for the moment a diagram which satisfies the following conditions (9) and (10).

(9) At each vertex V_j there exists exactly one external line that touches V_j. The external line shall be indexed to be L_j^e and supposed to be incoming.

(10) The diagram D is simple in the sense that for any pair of vertices (V_{j_1}, V_{j_2}) there exists at most one internal line (possibly no) that joins V_{j_1} and V_{j_2}.

Example of a diagram D *satisfying conditions* (9) *and* (10):

D:

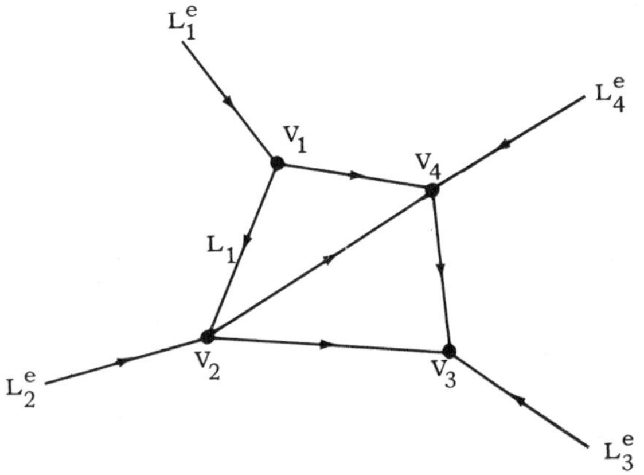

A (simple) daughter diagram D_1 of D with respect to L_1 is, by definition, a diagram obtained from D by deleting L_1 and identifying the end points W_1^+ and W_1^- of L_1.

The diagram D_1 *obtained from* D *in the preceding example:*

D_1:

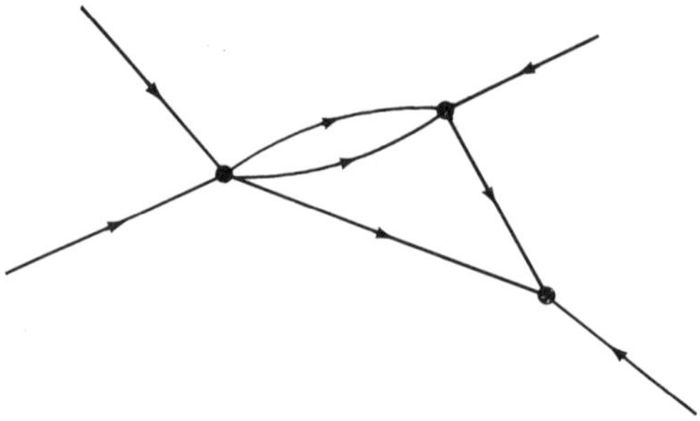

We denote by $\mathcal{L}_0(D^+)$ the part of $\mathcal{L}(D^+)$ where all $\alpha_\ell \gneq 0$. In our case we can verify that $\overline{\mathcal{L}_0(D^+)}$ and $\mathcal{L}_0(D_1^+)$ intersect normally along a codimension 1 submanifold. Hence, if we assume in addition that $\overline{\pi(\mathcal{L}_0(D^+))}$ (*) and $\pi(\mathcal{L}_0(D_1^+))$ are real hypersurfaces (i.e., with codimension 1) in $M = \left\{\sum_{j,r} [j:r]p_r = 0\right\}$, then we can introduce a local coordinate system on M near $p_0 \in \overline{\pi(\mathcal{L}_0(D^+))} \cap \pi(\mathcal{L}_0(D_1^+))$ so that $\overline{\pi(\mathcal{L}_0(D^+))} = \{x \in \mathbf{R}^{4n-4};\ x_1 = x_2^2,\ x_2 \geq 0\}$ (**) and $\pi(\mathcal{L}_0(D_1^+)) = \{x \in \mathbf{R}^{4n-4};\ x_1 = 0\}$. Then a result of Kashiwara-Kawai-Oshima [13] asserts the following:

THEOREM 3. *Under the assumptions on* D *so far stated,* $f_D(p)$ *has the following form* (11.A) *or* (11.B) *near the point* $p_0 \in \overline{\pi(\mathcal{L}_0(D^+))} \cap \pi(\mathcal{L}_0(D_1^+))$ *in the coordinate system introduced above. Here* $\phi_j(x)$ $(j = 1, 2, 3)$ *are holomorphic functions defined near* p_0 *and* $F(\alpha, \beta, \gamma; z)$ (***) *is the hypergeometric function.*

Case A: $\nu \equiv -2n' + (3N/2) + 3/2 \in \{0, 1, 2, \cdots\}$.

(*) π denotes the canonical projection from $\sqrt{-1}\, S^*M$ to M.

(**) The condition $x_2 \geq 0$ arises from the positivity of α_ℓ's. The part $\{x;\ x_1 = x_2^2,\ x_2 < 0\}$ corresponds to the solutions of Landau-Nakanishi equations with $\alpha_1 < 0$ and $\alpha_\ell > 0$ $(\ell \neq 1)$.

(***) We consider the analytically continued function outside $\{z \in \mathbf{C};\ z \in \mathbf{R},\ z \geq 1\}$.

$$\text{(11.A)}\quad \phi_1(x)\left(\frac{(-1)^{\nu+1}}{2\nu!}(x_1-x_2^2)^\nu \log(x_1-x_2^2+\sqrt{-1}\,0)\right.$$
$$\left.+\frac{\Gamma(-\nu+1/2)}{\sqrt{\pi}}x_2(x_1+\sqrt{-1}\,0)^{\nu-\frac{1}{2}}F\left(-\nu+1/2,1,3/2;\frac{x_2^2}{x_1+\sqrt{-1}\,0}\right)\right)$$
$$+\phi_2(x)\left(\frac{(-1)^{\nu+1}}{\nu!}x_2(x_1-x_2^2)^\nu \log(x_1-x_2^2+\sqrt{-1}\,0)\right.$$
$$\left.+\frac{\Gamma(-\nu-1/2)}{\sqrt{\pi}}(x_1+\sqrt{-1}\,0)^{\nu+\frac{1}{2}}F\left(-\nu-1/2,1,1/2;\frac{x_2^2}{x_1+\sqrt{-1}\,0}\right)\right)$$
$$+\phi_3(x)\,.$$

Case B: $\nu-1/2\in\{0,1,2,\cdots\}$.

$$\text{(11.B)}\quad \phi_1(x)\left(\frac{\Gamma(-\nu)}{2}(x_1-x_2^2+\sqrt{-1}\,0)^\nu+\frac{(-1)^{-\nu+\frac{1}{2}}}{\sqrt{\pi}(\nu-1/2)!}\right.$$
$$\times\left(-x_2x_1^{\nu-\frac{1}{2}}\log(x_1+\sqrt{-1}\,0)F\left(-\nu+1/2,1,3/2;\frac{x_2^2}{x_1+\sqrt{-1}\,0}\right)\right.$$
$$\left.\left.+x_2x_1^{\nu-\frac{1}{2}}\left(\frac{\partial}{\partial\alpha}F\left(\alpha,1,3/2;\frac{x_2^2}{x_1+\sqrt{-1}\,0}\right)\right)\Bigg|_{\alpha=-\nu+\frac{1}{2}}\right)\right)$$
$$+\phi_2(x)\left(\Gamma(-\nu)x_2(x_1-x_2^2+\sqrt{-1}\,0)^\nu+\frac{(-1)^{-\nu-\frac{1}{2}}}{\sqrt{\pi}(\nu+1/2)!}\right.$$
$$\times\left(-x_1^{\nu+\frac{1}{2}}\log(x_1+\sqrt{-1}\,0)F\left(-\nu-1/2,1,1/2;\frac{x_2^2}{x_1+\sqrt{-1}\,0}\right)\right.$$
$$\left.\left.+(x_1+\sqrt{-1}\,0)^{\nu+\frac{1}{2}}\left(\frac{\partial}{\partial\alpha}F\left(\alpha,1,1/2;\frac{x_2^2}{x_1+\sqrt{-1}\,0}\right)\right)\Bigg|_{\alpha=-\nu-\frac{1}{2}}\right)\right)$$
$$+\phi_3(x)\,.$$

See [13] Theorem 1.2, for the general formula which can be applied to generalized Feynman amplitude $f_D(p;\lambda)$.

We note that the geometrical situation discussed in Theorem 3, i.e., the situation where two positive-α Landau-Nakanishi varieties projected to p-space are osculating along codimension 1 submanifold, is also crucially important in discussing the relations among several Feynman amplitudes (e.g., relation between $f_D(p)$ and $f_{D_1}(p)$) –the so-called hierarchical principle. See [4], [10], [21], [27], [30] and references cited there for this topic.

We end this report by pointing out one interesting application of the theory of linear differential equations to the analysis of Feynman integrals, which differs in its nature from what has been discussed so far.

It is a problem of asymptotic behavior of Feynman integrals. One of the simplest problems of this sort is the study on the behavior of $F_D(p; \lambda; m)$ (*) as some masses m_ℓ ($\ell \in L_0 \subset \{1, \cdots, N\}$) tend to zero. In this case, at least for a diagram D satisfying the condition (9), we can explicitly find equations with regular singularities along $\{(p, m); m_\ell = 0, \ell \in L_0\}$ (not holonomic) which $F_D(p; \lambda; m)$ satisfies (Theorem 4 below). Then we use the results of Kashiwara-Oshima [15] to obtain the asymptotic behavior of $F_D(p; \lambda; m)$ as m_ℓ tends to zero (Theorem 5). This provides us with an alternative approach to the study of zero mass singularities of Feynman amplitudes done by Speer-Westwater [30] and Speer [29]. This topic shall be discussed in detail in a forthcoming paper of Jimbo-Kashiwara-Kawai-Oshima.

THEOREM 4. *The generalized Feynman integral* $F_D(p; \lambda; m)$ *satisfies the following differential equations with regular singularities along* $\{(p, m); m_\ell = 0\}$

$$(12) \quad \left[(m_\ell D_{m_\ell})^2 - m_\ell^2 \left(\sum_{j=1}^{n} [j:\ell] D_{p_j}\right)^2 + 2(\lambda_\ell - 2) m_\ell D_{m_\ell}\right] F_D = 0 \qquad (\ell = 1, \cdots, N)\ .$$

For simplicity let us consider the case where one m_ℓ, say m_1, tends to zero. Then we find the following

(*) In order to emphasize that $F_D(p; \lambda)$ depends also on m_ℓ's, we use this notation here.

THEOREM 5. *Suppose that* λ_1 *is not an integer. Let us denote by* $F_{D(m_1=0)}(p;\lambda;m')$ (*resp.* $F_{D'}(p;\lambda;m')$) *the integral obtained by replacing* $(K_1^2-c_1^2m_1^2+\sqrt{-1}0)^{-\lambda_1}$ *by* $(K_1^2+\sqrt{-1}0)^{-\lambda_1}$ (*resp.* $c_1^4\delta^4(K_1)$). *Here* $m'\equiv(m_2,\cdots,m_N)$. *Then the boundary value (in the sense of* [15] *Definition* 4.8) *of* $F_D(p;\lambda;m)$ *along* $\{(p,m);\ m_1=0\}$ *associated with the characteristic exponent*(*) 0 *is given by* $F_{D(m_1=0)}(p;\lambda;m')$ *and that associated with the characteristic exponent* $2(2-\lambda_1)$ *is given by* $e^{\lambda_1\pi\sqrt{-1}}\pi^2\lambda_1(\lambda_1+1)F_{D'}(p;\lambda;m')$. *Furthermore,* $F_D(p;\lambda;m)$ *has the following form* (13) *in a neighborhood of* $\Lambda^+\equiv\{(p,m;\sqrt{-1}\,dm_1\infty)\in\sqrt{-1}\,S^*R^{4n+N};\ m_1=0\}$:

$$F\left(\frac{1}{2},1,\lambda_1-1;\frac{\left(\sum_{j=1}^{n}[j:1]D_{p_j}\right)^2}{D_{m_1}^2}\right)F_{D(m_1=0)}Y(m_1)$$

(13)

$$+\left(1-\frac{\left(\sum_{j=1}^{n}[j:1]D_{p_j}\right)^2}{D_{m_1}^2}\right)^{\lambda_1-5/2}\left(e^{\lambda_1\pi\sqrt{-1}}\pi^2\lambda_1(\lambda_1+1)F_{D'}m_{1+}^{2(2-\lambda_1)}\right).$$

Note that the operators used in (13) are well-defined micro-differential operators defined near Λ^+.

REFERENCES

[1] Chandler, C., Some physical region mass shell properties of renormalized Feynman integrals, *Commun. math. Phys.*, *19* (1970), 169-188.

[2] Chandler, C. and H. Stapp, Macroscopic causality conditions and properties of scattering amplitudes, *J. Math. Phys.*, *10* (1969), 826-859.

(*) A characteristic exponent is, by definition, a solution of the indicial equations associated with (12). (See [15] p. 174.) In our case, the indicial equation is $s(s+2(\lambda_1-2))=0$.

[3] Chew, G., *The Analytic S-matrix*, Benjamin, New York, 1966.

[4] Eden, R., P. Landshoff, D. Olive and J. Polkinghorne, *The Analytic S-matrix*, Cambridge University Press, 1966.

[5] Iagolnitzer, D., Analyticity property of scattering amplitudes: a review of some recent developments, *Lecture Notes in Phys.*, 39, Springer-Verlag, Berlin-Heidelberg-New York, 1975, 1-21.

[6] ———, The structure theorem in S-matrix theory, *Commun. math. Phys.*, *41* (1975), 39-53.

[7] Iagolnitzer, D. and H. Stapp, Macroscopic causality and physical region analyticity in S-matrix theory, *Commun. math. Phys.*, *14* (1969), 15-55.

[8] Jimbo, M., A correction to "Holonomy structure of Landau singularities and Feynman integrals," *Publ. RIMS, Kyoto Univ.*, *12*, Suppl. (1977), 438-439.

[9] Kashiwara, M. and T. Kawai, Holonomic systems of linear differential equations and Feynman integrals, *Publ. RIMS. Kyoto Univ.*, *12*, Suppl. (1977), 131-140.

[10] ———, Holonomic character and local monodromy structure of Feynman integrals, *Commun. math. Phys.*, *54* (1977), 121-134.

[11] ———, On holonomic systems for $\prod_{\ell=1}^{N} (f_\ell + \sqrt{-1}\, 0)^{\lambda_\ell}$, to appear in *Publ. RIMS, Kyoto Univ.*

[12] ———, On holonomic systems of micro-differential equations III-systems with regular singularities, to appear.

[13] Kashiwara, M., T. Kawai and T. Oshima, A study of Feynman integrals by micro-differential equations, *Commun. math. Phys.*, *60* (1978), 97-130.

[14] Kashiwara, M., T. Kawai and H. Stapp, Micro-analytic structure of the S-matrix and related functions, *Publ. RIMS, Kyoto Univ.*, *12*, Suppl. (1977), 141-146. A full paper will appear in *Commun. math. Phys.*

[15] Kashiwara, M. and T. Oshima, Systems of differential equations with regular singularities and their boundary value problems, *Ann. of Math.*, *106* (1977), 145-200.

[16] Kawai, T. and H. Stapp, Micro-local study of the S-matrix singularity structure, *Lecture Notes in Phys.*, 39, Springer-Verlag, Berlin-Heidelberg-New York, 1975, 36-48.

[17] ———, Discontinuity formula and Sato's conjecture, *Publ. RIMS, Kyoto Univ.*, *12*, Suppl. (1977), 155-232.

[18] Landau, L. D., On analytic properties of vertex parts in quantum field theory, *Nucl. Phys.*, *13* (1959), 181-192.

[19] Nakanishi, N., Ordinary and anomalous thresholds in perturbation theory, *Prog. Theor. Phys.*, *22* (1959), 128-144.

[20] ————, *Graph Theory and Feynman Integrals*, Gordon and Breach, New York, 1971.

[21] Pham, F., Singularités des processus de diffusion multiple, *Ann. Inst. H. Poincaré*, *6A* (1967), 89-204.

[22] ————, Microanalyticité de la matrice S, *Lecture Notes in Math.*, 449, Springer-Verlag, Berlin-Heidelberg-New York, 1975, 83-101.

[23] Regge, T., Algebraic topology methods in the theory of Feynman relativistic amplitudes, Report of *Battele Renctres.*, Benjamin, New York, 1968, 433-458.

[24] ————, Old problems and new hopes in S-matrix theory, *Publ. RIMS, Kyoto Univ.*, *12*, Suppl. (1977), 367-375.

[25] Sato, M., Recent development in hyperfunction theory and its application to physics, *Lecture Notes in Phys.*, 39, Springer-Verlag, Berlin-Heidelberg-New York, 1975, 13-29.

[26] Sato, M., T. Kawai and M. Kashiwara, Microfunctions and pseudo-differential equations, *Lecture Notes in Math.*, 287, Springer-Verlag, Berlin-Heidelberg-New York, 1973, 265-529.

[27] Sato, M., T. Miwa, M. Jimbo and T. Oshima, Holonomy structure of Landau singularities and Feynman integrals, *Publ. RIMS, Kyoto Univ.*, *12*, Suppl. (1977), 387-438.

[28] Speer, E. R., *Generalized Feynman Amplitudes*, Princeton University Press, 1969.

[29] ————, Mass singularities of generic Feynman amplitudes, *Ann. Inst. H. Poincaré*, *26* (1977), 87-105.

[30] Speer, E. R. and M. J. Westwater, Generic Feynman amplitudes, *Ann. Inst. H. Poincaré*, *14* (1971), 1-55.

Library of Congress Cataloging in Publication Data

Guillemin, V 1937-
Seminar on micro-local analysis.

(Annals of mathematics study ; no. 93)
1. Mathematical analysis--Addresses, essays, lectures. I. Kashiwara, Masaki, 1947- joint author. II. Kawai, Takahiro, joint author. III. Title. IV. Series: Annals of mathematics studies ; no. 93.
QA300.5.G84 515 78-70609
ISBN 0-691-08228-6
ISBN 0-691-08232-4 pbk.